Praise for Zazie Todd and *Wag*

"Zazie Todd does Dogs the immeasurably good favor of taking their happiness seriously. Todd is dialed in to the science of dogs and a thoughtful trainer of dogs. Everything she writes about, you want to know. *Wag* is a welcome addition to the books geared to helping you help your dog."

ALEXANDRA HOROWITZ, author of *Our Dogs, Ourselves: The Story of a Singular Bond*, and *Inside of a Dog: What Dogs See, Smell, and Know*

"Practical, compassionate, thorough—and based on science rather than wishful thinking—*Wag* is a gift you should give to yourself and the dog or dogs in your life. I loved it."

CAT WARREN, author of *What the Dog Knows: Scent, Science, and the Amazing Ways Dogs Perceive the World*

"If you care about your dog, you need this book. It's packed with insights from the latest canine science, and loads of advice on how you can give your dog the happiest possible life."

DR. JOHN BRADSHAW, author of *Dog Sense: How the New Science of Dog Behavior Can Make You a Better Friend to Your Pet* (*In Defence of Dogs: Why Dogs Need Our Understanding* in the UK)

"*Wag* is a must-read for all dog lovers. An amazing, well-written book that makes canine science easy to understand and apply in a useful manner to improve dogs' lives and the relationship we have with them. The anecdotes about Dr. Todd's dogs are the highlight of the book."

DR. WAILANI SUNG, board-certified veterinary behaviorist, San Francisco SPCA

"Love dogs? Then you'll love *Wag*! Dr. Todd skillfully translates dozens of recent scientific studies into practical recommendations for all of us who strive to reciprocate the joy our dogs provide us so often and so well."

KATHY SDAO, applied animal behaviorist and author of *Plenty in Life Is Free: Reflections on Dogs, Training, and Finding Grace*

"In *Wag*, psychologist and dog trainer Zazie Todd shows dog owners how they can use the latest research in the exploding field of canine science to improve the lives of their pets. Dog lovers will find this book fascinating, and their pets will be delighted their owners have read it."

HAL HERZOG, author of *Some We Love, Some We Hate, Some We Eat: Why It's So Hard to Think Straight about Animals*

"Beautifully written and meticulously researched, *Wag* brings the latest and best science on dog welfare, behavior, training, and health to bear on the paramount topic for all of us interested in dogs: their happiness. Required reading."

JEAN DONALDSON, the Academy for Dog Trainers

"I have long admired Dr. Todd's ability to smoothly blend current science into her incredibly informative, engaging, useful, relevant, dog-related writing, and this book is no exception. I will be making it required reading for all my dog trainer academy students and interns, adding it to my recommended reading list for all of my clients, and I also suggest that it be mandatory reading for all who care about making life happier—and better—for dogs."

PAT MILLER, Peaceable Paws and training editor for *Whole Dog Journal*

"Our species has spent thousands of years trying to shape domestic dogs to our ever-changing lifestyles, but only in the past century have we seriously considered the unique needs of our canine companions. Fortunately for all of us, Dr. Zazie Todd has written a delightful, compact, evidence-based guide to building better partnerships with our dogs and sending more joy their way."

BRONWEN DICKEY, author of *Pit Bull: The Battle over an American Icon*

"*Wag: The Science of Making Your Dog Happy* is one of the most comprehensive and engaging books about dog behavior that you will ever read."

VICTORIA STILWELL, celebrity dog trainer and host of *It's Me or the Dog*

"The more science learns about dogs, the more extraordinary they turn out to be. And the more we understand them, the better able we are to care for them, to guide them, to make them happy. But cutting-edge science can be hard to get your head around. So thank you Zazie Todd for sorting through it all, making it relevant, and keeping it simple. This is a book your dog will want you to read."

DR. MARK EVANS, former RSPCA chief vet and TV presenter

"Well-written and packed with great advice, this book could fundamentally change the relationship between you and your dog."

DAVID GRIMM, author of *Citizen Canine: Our Evolving Relationship with Cats and Dogs*

THE SCIENCE
OF MAKING YOUR
DOG HAPPY

Foreword by **DR. MARTY BECKER**

ZAZIE TODD

GREYSTONE BOOKS
Vancouver/Berkeley

Greystone Books Ltd.
greystonebooks.com

Cataloguing data available from Library and Archives Canada
ISBN 978-1-77164-379-5 (pbk.)
ISBN 978-1-77164-380-1 (epub)

Editing by Lucy Kenward
Copyediting by Rowena Rae
Proofreading by Jennifer Stewart
Cover and text design by Belle Wuthrich
Cover photograph by GalapagosPhoto/Shutterstock.com

Printed and bound in Canada on ancient-forest-friendly paper by Friesens

Greystone Books gratefully acknowledges the Musqueam, Squamish, and Tsleil-Waututh peoples on whose land our office is located.

Greystone Books thanks the Canada Council for the Arts, the British Columbia Arts Council, the Province of British Columbia through the Book Publishing Tax Credit, and the Government of Canada for supporting our publishing activities.

Canadä

BRITISH COLUMBIA

BRITISH COLUMBIA ARTS COUNCIL
An agency of the Province of British Columbia

Canada Council   Conseil des arts
for the Arts   du Canada

For Ghost
And also for Bodger

CONTENTS

FOREWORD

M
Y LOVE OF animals first led me to become a veterinarian, and I have had the great fortune to spend my life advocating for animal welfare and meeting many others who do too. I have authored twenty-five books about pets, was the resident veterinarian on *Good Morning America* for seventeen years, and have had a nationally syndicated pet/vet column for almost twenty years. I'm also the founder of Fear Free, an initiative whose goal is to make visits to the veterinarian less stressful for pets, so I know the importance of teaching pet owners how to care for their pets' physical and emotional health. That's why I'm so thrilled to see this book by Dr. Zazie Todd.

As a regular reader of Todd's fascinating blog, *Companion Animal Psychology*, I know that she loves nothing more than writing about dogs, science, and happiness. And in this book, Todd clearly and expertly explains the science that tells us how to ensure our dogs are happy, elaborating on many of the subjects she covers in her blogposts and touching on lots of new ones as well. *Wag* begins by explaining how and why we should think about good welfare for our dogs, and then takes us on a tour of what science tells us about every aspect of a pet dog's life. From making good decisions when choosing a puppy to getting socialization right, from the importance of play to the best ways to provide enrichment to engage dogs' brains, and how to have safe, happy interactions with people of all ages, *Wag* has it all. Each chapter ends with user-friendly tips that pet owners can apply in their everyday lives.

With Todd's background in psychology, the importance of emotional well-being is something she knows only too well. When people know how to increase positive welfare for their pets, it makes a difference to their dog—regarding both physical and emotional health—and it improves the human-animal bond. *Wag* gives dog owners the knowledge and the tools to augment their dog's positive welfare with a goal of optimal happiness and enrichment.

When she interviewed me in 2018 about my most recent book, *From Fearful to Fear Free: A Positive Program to Free Your Dog from Anxiety, Fears, and Phobias*, I gave her the authentic story behind Fear Free, and my stunning realization that failing to take care of the emotional health of pets can cause them physical harm. Increasing feelings of happiness and calmness and providing plenty of enrichment is the key to good animal welfare (and also an integral part of Fear Free Happy Homes, which helps people enrich their pets' lives). Research about animal welfare and behavior is a fast-changing field, and this important book is meticulously researched and up to date on the latest science about dogs. In addition, Todd has interviewed canine scientists, veterinarians, veterinary behaviorists, shelter managers, and dog trainers to find out what they most want people to know about dogs. Whether they are talking about their research from a dog's point of view or sharing the one tip they think would make the world better for dogs, their perspectives are enlightening and sometimes surprising.

Todd is not only an engaging and encouraging science writer who is passionate about good animal welfare, she works with dogs too. She is a member of the Fear Free Advisory Group, a certified dog trainer, and the owner of several pets. As a result, *Wag* is filled with lots of relatable personal anecdotes about Todd's own dogs, Bodger the Australian Shepherd and Ghost, a Siberian Husky/Alaskan Malamute cross. We learn about how they came into her life, challenges during their training sessions, and their doggy quirks. Todd's love of dogs shines through in these charming stories.

If, like me, you care about dogs' emotional welfare, you will want to read this book. *Wag* is scientifically accurate and beautifully written, a rare blend of science and soul. Read on to find out how *Wag* can help you have an even happier dog.

—MARTY BECKER, DVM

INTRODUCTION

FALLING FOR DOGS

EVERY DOG CHANGES your life to a greater or lesser degree. I never imagined a beautiful Siberian Husky/Alaskan Malamute cross would change mine so significantly, but that's how it turned out.

It was a hot day when we drove Ghost home. We let him out of the car in the garage and took him outside for a quick on-leash toilet break before bringing him in. I thought he seemed relieved to arrive at a house rather than a kennel; he had, after all, already been transferred from one animal control to another. He was 4 years old, or at least that's what we were told, though he grew a little in his first few months with us. He was understandably nervous at arriving in a new place with people he did not yet know, and I spent a lot of those early days petting him. He would lie on his side, sprawled across the room, and I would kneel down and stroke him, his fur soft and thick. However long this went on, when I stopped he would raise his big head, lick his lips, and sometimes paw at me for more. But although he liked petting, in other circumstances he would shy away from our hands as if we were about to hurt him.

Everywhere we went, people complimented his looks. I could not believe such an amazing dog had waited so many weeks for someone to adopt him. He was friendly, handsome, and took up a lot of space. He would not make eye contact but I had the feeling, whenever my back was turned, that he was watching me. He had beautiful pale blue eyes, almost white, ringed with black eyeliner. When he moved, he left little bits of fur behind; shedding continued all summer and whenever I brushed him, I seemed to remove enough fur to make a whole other dog. He wasn't quite sure what to do and nor were we. How much to feed him, how many walks a day, what kinds of games he would like—all of this we had to find out. He wasn't interested in people we met on walks, though he tolerated their patting, but he loved meeting other dogs. Several times I felt my arm yanked as he pulled to sniff another dog. Pretty soon we decided he needed a friend.

An Australian Shepherd was listed at our local shelter, without an online photo or any information yet. This seemed to be a trainable breed according to my research on the internet, so we took Ghost to meet him. The dog was at the groomer, and we waited on chairs in reception for him to come back. Ghost lounged patiently on the floor. The Aussie's return was delayed, and we came home only to get a phone call saying we should go right back. It was almost closing time by then, and we took the Aussie, beautifully groomed and in a bandanna, for a very short walk. Ghost seemed to like him, so that was that. We called him Bodger. In the space of six weeks, we had gone from no dogs to two dogs. (Oh, and two cats, because while we waited to meet Bodger, a tabby cat kept looking at Ghost through the glass, so we brought him home too.

And since he needed a friend, he was quickly followed by a pretty tortoiseshell.)

I kept being puzzled by things I read about dogs or saw on TV. Your dog was not supposed to go in front of you on a walk, except Ghost was a sled dog and they had to be in front of a sled or they couldn't pull it. You had to eat before your dog, but that wasn't convenient as it suited our routine to feed the animals first. You should be able to take things like bones out of your dog's mouth. Well, that just seemed stupid! Ghost had big teeth and I wasn't about to risk finding out what they might do. Besides which, he was agreeable when it came to swapping things. If I wanted the ball back, he would gladly trade it for some treats. Why be confrontational when you don't need to be?

I'll be honest: Bodger was a bit nuts. He was jumpy and mouthy, barked nonstop, and seemed to pull in all directions at once on the leash. He constantly nudged us for attention but was quick to growl when we looked at him. Left to his own devices for a moment, he would grab his tail in his mouth and spin in never-ending circles. Some of the people who met him in those first few weeks assumed we would take him straight back to the shelter. It might have been tempting, except we felt we had a responsibility towards him now. He hadn't been well socialized, apparently, and no one had ever taught him what to do. We just had to teach him. At least he was house-trained. And he seemed to understand the bit where his job was to keep Ghost company. Life was suddenly about keeping both dogs happy, which was not as easy as you might think.

I would like to say those myths about eating before your dog and so on that I saw on TV have faded away, but they are still widely believed. And at the same time, what we know about dogs—really know, that is—has dramatically increased. Even though there's a lot still unknown, we have a better understanding of dogs than ever before.

LEARNING ABOUT DOGS

WHEN I DID my PhD in Social Psychology at the University of Nottingham, I taught a range of topics in basic psychology. I even helped students dissect sheep's brains, teaching them to identify the different parts such as the hippocampus (important for memory and emotions) and the olfactory bulb (important for smelling, as you might guess from the name). I still remember the smell of preservative and the grayish-yellowish color of the brains. And I taught Psychology 101 tutorials with first-year undergraduates on topics such as how animals (including humans) learn.

Six students at a time would crowd into my office for their tutorial, a small participatory class that is a standard in the British university system. For the topic of animal learning, I asked students to come with examples of types of reinforcement and punishment. One of my own examples was about my ginger-and-white cat, Snap. At night when it was time for him to come in, I would call him and shake the treat packet. When he came in the kitchen door, I would give him a treat. This is an example of positive reinforcement because giving a treat made him more likely to come when called the next night. Although most students had thought of human examples, some had examples from their family dog. This is such useful information for anyone who has a pet. It's one of many reasons I wish everyone studied some basic psychology, because when we understand the rules of behavior, it makes for a happier relationship with your dog.

Aside from occasional tutorials or conference talks I attended, I didn't think much about animal learning until we adopted Ghost.

It was then I was confronted both with the reality of having a real live dog to care for and the difficulties of finding good advice. In 2012, less than a year after adopting Ghost, I started my blog *Companion Animal Psychology* with the aim of finding out more about what science tells us about how to care for dogs and cats. I found a rich vein of canine and feline science to write about, and many people eager to learn more. The burgeoning field of canine science means there is something new to learn, even for lifelong dog people.

On the one hand, those principles of reinforcement and punishment are central to how we live with pets, and on the other hand, what we know about animals' thoughts and feelings has been completely transformed. From thinking of animals as simply responding to stimuli, as American psychologist Burrhus Frederic (B.F.) Skinner believed, we now recognize animals as sentient beings. We know our pets have thoughts and feelings, including about us. And that means we have an even greater responsibility to take care of them in ways that recognize them for what they are—clever beings who become attached to us and have complex needs of their own.

One of the findings that piqued scientists' interest in dogs came in 1998 with the simultaneous but separate discovery by Dr. Brian Hare and Prof. Ádám Miklósi that dogs can follow pointing gestures—something that chimpanzees, our closest relatives, cannot. Since then, research on dogs—their behaviors, emotions, responses to humans, etc.—has flourished. And not just on dogs, but all kinds of animals. Back when I helped students dissect sheep brains, humans and other animals were seen as far apart, with many abilities attributed to humans alone. It's as if, over time, the gap between humans and other animals has shrunk substantially (of course, it's just our perception that has changed).

For dog lovers, it's hard to imagine that scientists used to think animals did not experience emotions. But now scientists are interested in everything dog, from the origins of domestication to the "guilty look," from puppy development to play behavior. One of the great things about canine science is that so much of it has implications for animal welfare and how we can best take care of our dogs.

The same year I started my blog, I began to volunteer with my local branch of the British Columbia Society for the Prevention of Cruelty to Animals (BC SPCA), one of the leading SPCAs in North America. I wanted more experience with dogs and cats, and felt grateful to them since they were the source of three of my pets. A year later, I was lucky enough to win a scholarship to Jean Donaldson's prestigious Academy for Dog Trainers, a science-based course that teaches fast, efficient dog training as well as behavior modification for issues like fear, food guarding, and aggression. And finally I set up my business, Blue Mountain Animal Behaviour, to help dog and cat owners resolve their pets' issues. All the while, I've posted to my blog almost every Wednesday and started a second blog on *Psychology Today*.

If you had told my younger self that one day I would be writing about the science of how to make dogs happy, I would have been very surprised. Like many people, in the past I underestimated dogs. I'm not the only person whose dogs have prompted them to learn more about training and behavior, so this book is for everyone who wants to know more. I'm lucky to be in a position to understand (and contribute to) the science, and to have worked with all kinds of dogs. I love seeing the difference it can make to both human and dog when an owner has a better understanding of their dog's needs.

This book is about what science tells us about dogs and what it means for their welfare. Different chapters look at getting a dog, how to train your pet, the social behavior of dogs and how to tell when they are playing, what dogs eat, how much they sleep, and how to make visits to the veterinarian easier. There's even a chapter on end-of-life issues for those people who are struggling with this difficult time (ideally, something to read long before you need to think about this). Although I write about specific scientific studies, I've tried to make them easy to understand and not get bogged down in technical details. The book also includes many quotes from experts who answered my question, What's the one thing that would make the world better for dogs?

Every chapter ends with a set of bullet points telling you how to apply the science at home. They are realistic and evidence-based. There's a checklist at the end of the book to help you think about how to apply these ideas. The final chapter summarizes the most important things you can do for your dog.

By the end of the book, you'll have a good understanding of how to make your dog happy (or even happier). Of course, a book is not a substitute for a professional opinion. If you have concerns about your dog, see your veterinarian, dog trainer, or behaviorist, as appropriate.

And remember, we are always learning. Whatever we thought we knew about dogs is subject to change—and, as you'll see from this book, some of those developments are exciting, surprising, and relevant to our everyday lives. Let's start by looking at the things that need to come together to have a dog who is happy, not just in the moment but throughout life.

1

HAPPY DOGS

·················

GHOST LOVED THE snow and his thick fur coat was built for it. He would bound, jump, and roll in it, eat the fresh snow, and carefully sniff the yellow snow, nose twitching delicately as he took in every detail. There's a photo of him lying in deep snow in our backyard, looking long and lean as always. He's staring at the camera and his mouth is closed as if to say, "Why are you pointing that thing at me?" But either side of that moment, when the camera was away, he was ecstatic in his element.

Bodger loves to chase snowballs. When I kick snow into the air he tries to catch it, and as the excitement builds he goes *boing! boing! boing!* just watching my feet crunch on the snow. And at any time of year, he loves to be chased, especially if he has a stick. He will let me get quite close and then suddenly bound away across the lawn, stick firmly in his mouth as he does a victory lap before settling down to tempt me to come close again.

But happiness isn't just about these moments of delight; it's also about everyday contentment. There are several ingredients to having a happy dog: a happy dog must have their welfare needs met, which can only be done when we have a good knowledge of canine behavior and an understanding of what our individual dog needs; a happy dog of course must be happy, something we need to be able to recognize; and a happy dog must have a good relationship with their owner, because otherwise they are at risk of being rehomed or euthanized.

People want their dogs to be happy. We spend more on our pets than ever before. The American Pet Products Association estimates that Americans will spend more than $75 billion on pets in 2019 (a huge increase from twenty years ago when the amount was only $23 billion).[1] It is estimated that there are 89.7 million dogs in the United States, 8.2 million in Canada, and 9 million in the United Kingdom.[2] That's a lot of dogs to keep happy.

SPOTTING A HAPPY DOG

IT'S EASY TO spot a happy dog in the moment. The eyes are relaxed and the mouth is open in a relaxed way. Some teeth and part of the tongue are visible, but the lips are not pulled back to show off all the teeth in a snarl. Maybe the tail is wagging a lovely, loose wag that makes the whole body wiggle. The posture is normal, not lowered in fear, and the ears are relaxed.

Recognizing fear in dogs is harder, something people with professional experience are better at than regular dog owners.[3] Even in situations where people might reasonably expect their dog to be afraid, such as at the vet or when there are fireworks, a sizeable number of people miss the signs.[4] There are many ways dogs telegraph fear, anxiety, and stress: tucking the tail, holding the ears back, licking the lips or nose, making whale eyes (wide eyes showing the whites of the eyes), looking away, lifting a paw, trembling or shaking, having a low body posture, yawning, panting, grooming, sniffing, seeking out people (looking for comfort from their owner), hiding, not moving (often mistaken for being

This dog's relaxed eyes and open mouth show she is happy. *BAD MONKEY PHOTOGRAPHY*

Gemma does not like the camera, so she looks away. *CHRISTINE MICHAUD*

Although you can see teeth, the mouth is open in a relaxed way. *BAD MONKEY PHOTOGRAPHY*

Signs of stress. The dog is looking away, the mouth is closed, you can see whale eye, and the ears are pinned back. *KRISTY FRANCIS*

calm), having a stiff or frozen posture, urinating, and defecating. When people fail to spot these signs, they are not able to help their dog be less stressed.

Not all wags are friendly; a short, rapid wag with a high tail is a threat signal. However, some dogs are bred to have only a stubby or corkscrew tail, while cosmetic procedures are sometimes used to dock the tails and/or crop the ears. These breeding and cosmetic changes can interfere with our (and other dogs') ability to read canine body language. Some jurisdictions, such as British Columbia and Nova Scotia in Canada, have banned ear cropping and tail docking, but they are still permissible in many locations. Even other dogs get confused by stubby tails. When researchers made a robot dog that could have either a short stubby tail or a long (normal) tail, they found the tail made a difference to how other dogs behaved.[5] With the long tail, other dogs approached the robot when it wagged in a friendly way and stayed away when its tail was still and upright (a threat signal). But when the robot dog had a stubby tail, dogs approached cautiously as if they were not sure whether or not its intentions were friendly, regardless of what the tail did.

We take it for granted that dogs experience happiness and fear. Charles Darwin believed that human and non-human animals evolved the ability to experience emotions, but over the years many scientists have been skeptical, in part due to our inability to know the subjective experience of animals (and perhaps also because of historical beliefs about humans being unique and special compared with other animals).[6] But increasingly we have evidence of non-human animals experiencing emotion, and scientists are placing a greater emphasis on researching positive emotions instead of only negative ones like pain. And this means emotion needs to be part of our models of animal welfare.

The late neuroscientist Prof. Jaak Panksepp—perhaps best known for his research on tickling rats—identified seven primary emotion systems in the brain of animals (and people).[7] Four of these are positive: SEEKING (includes curiosity, anticipation, and enthusiasm), PLAY, LUST, and CARE (such as taking care of young). The other three systems are negative: RAGE (anger), FEAR, and PANIC (loneliness or sadness). They are written in capital letters because they refer to specific systems in the brain, not to the everyday sense of the words. In case you're wondering about the rat tickling, it involves the PLAY system. Panksepp's research in affective neuroscience shows that we have to take the idea of animal consciousness and animal emotions seriously.

GOOD ANIMAL WELFARE

EXCITING DEVELOPMENTS IN animal welfare science apply to our pet dogs. Since the 1960s, animal welfare has been framed in terms of preventing cruelty. The framework for how we think about dogs' welfare comes from the Five Freedoms, proposed in the UK's 1965 Brambell Report on farm animal welfare.[8] The phrase was borrowed from a 1941 speech by Franklin D. Roosevelt who referred to Four Freedoms for US citizens. The Five Freedoms (see text box) were originally designed for farm animals and are seen as applying to companion animals too.[9]

The Five Freedoms

- Freedom from thirst, hunger, and malnutrition—by ready access to a diet to maintain full health and vigor.

- Freedom from thermal and physical discomfort—by providing a suitable environment including shelter and a comfortable resting area.

- Freedom from pain, injury, and disease—by prevention or rapid diagnosis and treatment.

- Freedom from fear and distress—by ensuring conditions that avoid mental suffering.

- Freedom to express (most) normal behaviors—by providing sufficient space, proper facilities, and company of the animal's own kind.

Of these five things, the freedom to express most normal behaviors is the least well known. In a British survey, only 18 percent of people recognized it as a welfare need.[10] The other four needs were identified by a majority, and only 4 percent of pet owners said they were not interested in knowing more about how to provide good animal welfare.

More recently, the Five Domains Model (see figure) was proposed by Prof. David Mellor of Massey University in New Zealand.[11] The two approaches are complementary. One of the key differences is the idea that we should not just think about preventing harm, but also about providing good experiences. In other words, for good welfare, animals (including pet dogs) should get to do things that make them happy.

Overview of the Five Domains Model

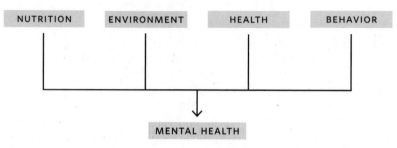

Source: Based on Mellor (2017)[12]

Prof. Mellor told me, "If you're talking about good nutrition, good environment, good health and appropriate behavior, what we need to make a distinction between is what we need in order to get animals to survive, and what we need in order not just to have them survive but to have them thrive."

Negative states cannot be removed entirely, Mellor said. Take thirst: without feeling thirst, animals (including us) would not drink; as we drink, the sensation of thirst goes away and we are no longer motivated to find water. Similarly, without hunger, animals would not eat. Although we cannot remove these experiences entirely, we can minimize them, and we can create positive experiences, for example with different types of food.

There's another kind of negative internal state to consider. The animal's perception of its environment and what is happening in it may cause negative emotions like fear, anxiety, depression, boredom, and loneliness. We are often responsible for the situations that cause these emotions, but that also means we can change them, for example by enriching the environment to prevent boredom. This is, Mellor said, "where we can have quite a profound influence on whether or not the animals can have positive experiences."

Mellor told me about the behavioral opportunities that dogs like to have. "We control a lot of these things," he said, "but it doesn't mean that a dog, to have a contented and happy and fulfilling life, needs to have access to all of those positive experiences. But the more that the dog can be given, appropriate to the circumstances, the better its life is going to be."

Negative welfare states such as fear or pain can stop dogs from experiencing positive states. For example, a dog in pain will not play, may withdraw from other animals and people, and may not eat. This is why it's important to minimize negative states as much as possible, not just in and of themselves, but also so the dog can experience pleasures. "And how do we know that they may be having positive experiences?" said Mellor. "It's because they are engaging in the behaviors that those opportunities enable them to do."

So to have a happy dog, we need to provide good nutrition, good health, a good environment, companionship, the ability to express appropriate behavior, and opportunities to experience positive emotions. A sense of well-being is not just about psychological welfare. Amongst orangutans living in zoos, those considered by their keepers to be "happy" lived longer lives.[13] And for both captive brown capuchins and chimpanzees, their keepers' ratings of their subjective "happiness" tie in with assessments of their positive and negative welfare.[14] While we don't have the same studies for dogs, we know the converse is true: stressed dogs live shorter lives.[15] So making our dogs happy may help them have longer, healthier lives. In the intricate balance between physical and emotional wellness, anything we can do to improve welfare may bring added benefits.

There are many welfare problems for pet dogs: confrontational dog training methods that risk fear, stress, and aggression; breeding practices that reduce genetic diversity and increase the risks of inherited disease; changes in working lives and living spaces that mean dogs may be left home alone for longer and have to meet many other dogs when on walks; tail docking, ear cropping, and debarking (where these procedures are legal) that cause pain and reduce communicative abilities; and people's failure to recognize signs of fear, anxiety, and stress in their pooch—or even people finding these signs funny. Some of these issues are caused by a failure to understand dogs.

UNDERSTANDING CANINE BEHAVIOR

UNDERSTANDING PET DOGS—WHO they are and why they behave the way they do—is also central to giving dogs a happy life. This idea was highlighted by Dr. Sam Gaines, head of the Companion Animals Department at the Royal Society for the Prevention of Cruelty to Animals (RSPCA) in the UK. "A lot of the problems that we see or hear about," she said, "wouldn't necessarily come to light if people had a much better understanding of the dog that they've actually got in their house. So, for example, people go and impulse-purchase a puppy without doing any research, and then suddenly end up with this little creature in their house which they have no or very little understanding about, which then means it's very difficult for them to provide for their welfare needs."

And unfortunately there is a lot of misinformation too, which means people's folk knowledge about dogs is often wrong. Gaines said, "In an ideal world what I would really like to do . . . is sort of like wipe the slate clean when it comes to [what people know about] dogs. Like in *Men in Black* they press that pen and every memory or anything associated disappears, and you can then give them a new knowledge and understanding of what a dog is."

One of the great things about canine science is that researchers are investigating topics that are important for the everyday lives of dogs. However long you have known dogs, there is something new and exciting to learn.

THE NEEDS OF THE INDIVIDUAL DOG

JUST LIKE PEOPLE, every dog is an individual. Some dogs are sociable and friendly; they love to meet new people and other dogs, and so we should try to give them more of these experiences. On the other hand, some dogs are shy and timid and would hate to be forced to meet other people and dogs every day. That's okay, because the important thing is that we recognize the needs of the dog we have and cater to them.

Individual differences were apparent with Ghost and Bodger. While Ghost was calm, sometimes aloof with other people, Bodger is desperate to become their friend. Having learned that sitting is required before he is patted, he secretly waits for just the right moment to leap up and lick the unsuspecting person on the face. And while Ghost was always so happy to meet other dogs, Bodger is choosy about who is allowed in his space.

There are two sides to considering the needs of an individual dog. The first is to do with minimizing experiences the dog finds negative, such as preventing situations where the dog is fearful (which may include avoiding the situation, teaching the dog to like the situation instead, and/or using medication under the guidance of a veterinarian). The second is to do with knowing what that particular dog enjoys. Do they love to play fetch or do they prefer to go for a swim? Do they love agility class or do they prefer to mooch about on a forest trail? It's up to us to know what our dog likes and give them the opportunities to experience it.

THE IMPORTANCE OF THE HUMAN-ANIMAL BOND

WHEN WE GET a dog, we imagine a long, beautiful friendship, kind of like a canine equivalent of walking off into the sunset to live happily ever after. But although we think of dogs as our best friends, the relationship often breaks down. We know that:

- The American Society for the Prevention of Cruelty to Animals (ASPCA) says 670,000 dogs are euthanized every year in American shelters because they do not have a home.[16]

- The American Veterinary Society of Animal Behavior says behavior problems are the leading cause of death in dogs under 3 years old in the US.[17] In the UK, behavioral issues are responsible for 14.7 percent of deaths in dogs under three (compared with 14.5% from gastrointestinal issues and 12.7% from accidents involving cars) according to the *Veterinary Journal*.[18]

- The American Humane Association found 10 percent of newly adopted dogs and cats in the US are no longer in the new home six months later (either returned to the shelter, lost, dead, or given to someone else), while the BBC reports 19 percent of people in the UK who buy a puppy no longer have them two years later.[19]

Clearly, for many people who start out with high hopes for a relationship with a dog, things go badly wrong. In part, this may be due to a lack of preparation. Between 18 and 39 percent of dog owners do no research at all before getting a dog.[20] Of course other issues, such as a lack of pet-friendly rental housing or people becoming sick and no longer being able to care for their pet, may also play a role. Helping to prevent relationships with our pets from breaking down will make us happier as well as our dogs.

I think we all want to make our dogs happy, even if along the way we show it in different ways and sometimes do the wrong thing. We love to see a happy look on our dog's face, and let's face it, the bounding, bouncing joy of a dog is enough to make us happy too. As guardians, we are responsible for everything in our dog's life, and it's an understatement to say we are important from our dog's perspective. So this book is not just about your dog—it's about you and your dog, the human-canine partnership, and what canine happiness means.

"THINK DOG! DESPITE a wealth of research into the domestic dog and a greater understanding of how they behave, think, feel, and interact with us and their peers, many owners/guardians continue to treat dogs either as wolves or little people and/or fail to understand and acknowledge what dogs actually are. This can have a huge impact on their physical and mental health. For example, decades of thinking of dogs as wolves has contributed to a widespread use of management and training techniques that place dogs at serious risk of poor welfare. Similarly our failure to understand what it is to be a dog and what constitutes normal behavior can mean a poor quality of life through a lack of outlets for strongly motivated behavior such as playing, sniffing, and investigating. If dogs really are our best friends and we want them to be truly happy, then we have to think dog."

—SAM GAINES, PhD, head, Companion Animals Department, RSPCA

2

GETTING A DOG

....

WHEN I WAS in my thirties, I dreamed of getting a dog. At the time I wasn't home enough to care for a dog properly, so I dreamed of the dog I would get when my lifestyle changed. We would take long walks together through the countryside, and then I'd curl up on the settee with my dog and a good book.

I decided early on that my ideal dog would be like Diefenbaker in the TV show *Due South*, which was popular in the UK where I was living. On TV, Dief was a beautiful half-wolf, half-sled-dog who was loyal, independent, and deaf (or maybe only when he didn't want to hear?). Over the years, Diefenbaker was played by six different Siberian Huskies. I did some research on the breed: "Not for first-time dog owners," I kept reading, "escape artists," "independent," and "difficult." Not to mention the shedding. Although some of this information was off-putting, I figured I could cope.

Of course, I didn't get Diefenbaker; he only existed on TV. I was incredibly lucky to get Ghost instead. And I was not unusual in being influenced by TV and in thinking so much about the dog's appearance rather than other factors. It turns out many people are like this—and unfortunately this is often to the detriment of dogs. What I know now is that if we want a happy dog, there are many factors to take into account when choosing what kind of dog and where to get them. But many people are influenced by biology or fashion.

BIOLOGY AND LOVING DOGS

WE KNOW DOGS are the descendants of wolves, but in the process of domestication they have changed in appearance. Now, whatever your preference for looks, there is a breed to fit (even if the original purpose of the breed was for work). Some of the physical features of dogs are puppy-like rather than wolf-like, which may tap into our natural desire to help baby-like creatures. Is it just an accident that dogs evolved these different features, and does it affect the way we feel about them?

The Russian fox experiment is a pioneering experiment into the process of domestication that began in what was then the Soviet Union and continues today.[1] Geneticist Dmitri Belyaev had an idea that selecting animals for tameness would also lead to hormonal and other changes. He began a program to breed silver foxes. Only the tamest of each generation were selected for breeding, and crucially, nothing else was changed, so this was a test purely of genetics and not of handling or other factors. A second line of foxes was bred by choosing the most aggressive animals. During the first seven or eight months until they reached sexual maturity, the foxes were tested to see how they responded to the experimenters. Then, a choice was made as to which ones would be bred for the next generation.

Over time, as the foxes became more tame, other changes also happened. Prof. Lee Dugatkin, evolutionary biologist and co-author with Lyudmila Trut of a book about the study, *How to Tame a Fox (and Build a Dog): Visionary Scientists and a Siberian Tale of Jump-Started Evolution*, told me about the changes: "The only thing [the scientists] ever do to determine who is going to be the parents of the next generation in the experiment is test them on their behavior towards humans. That's it, that's the only thing they ever select on. But what's happened over the generations is that lots of other changes have occurred besides getting calmer and tamer animals. Early on, for example, some of the first changes were that the animals had curlier, bushier tails, the sort of tails that you imagine when you think of a dog wagging their tail because they're excited to see you. Some of the animals began to show droopier, floppier ears. In addition, they began to see a much more mutt-like kind of mottled fur color." There were also differences in stress hormones that showed the foxes were less stressed.

It is possible that some of the features we see in dogs are also by-products of domestication. However, another possibility is that along the way, we have sometimes selected for some of these features. To test this idea, scientists looked at one of the baby-like features in many dogs, the eyes. The study, published in PLOS ONE, looked at a facial expression by which dogs raise the inside of the eyebrow, making the eyes look bigger.[2] The researchers enlisted the help of four dog rehoming centers, and filmed the dogs for two minutes with an experimenter standing by the kennel. They counted how many times the dogs made this expression during that time. Then they waited to see how long it took for the dogs to be adopted. The results showed that dogs who made this eyebrow movement five times within the two-minute period were adopted in fifty days, compared with thirty-five days if they did it ten times, and twenty-eight days if they did it fifteen times. It seems the baby-like eyebrow movement results in people being more drawn to those dogs. This was the first time scientists demonstrated a link between the baby-like features of a dog and people's active selection of a dog.

HOW FASHION AFFECTS THE DOGS WE CHOOSE

BIOLOGY IS ONLY part of the story. Fashion affects breed popularity too. Featuring a particular breed of dog in the movies can increase its popularity for up to ten years afterward, according to a study in PLOS ONE that looked at dog-related movies from 1927 to 2004 and corresponding Kennel Club registrations.[3] The release of films such as 101 Dalmatians and The Shaggy Dog were followed by huge increases in the popularity of Dalmatians and Old English Sheepdogs, respectively. And the "movie effect" still holds even if the breed had been decreasing in general popularity before the film was released.

A review of dog breed registrations from 1926 to 2005, published in PLOS ONE, found the popularity of a breed is not affected by the breed's health, a longer lifespan, or better behavior (such as trainability, fearfulness, or aggression).[4] In other words, pet-keeping does not just have a biological explanation but is also socially mediated, according to a study in Animal Behavior and Cognition.[5] Just appearing in the media does not guarantee a breed's popularity, however. For example, winning Best in Show at the prestigious and televised Westminster Kennel Club Dog Show does not usually have an effect, as reported in the Journal of the American Veterinary Medical Association.[6]

Dr. Hal Herzog is a professor emeritus of psychology and author of Some We Love, Some We Hate, Some We Eat: Why It's So

Hard to Think Straight about Animals who has spent over three decades investigating our interactions with animals, and who was involved in this research on the popularity of dog breeds. He told me this research had a profound effect on his own views about the role of biology versus culture. "For many, many years I've considered myself an evolutionary psychologist and I still do, but I really strongly believed that most of our behavior was determined by biological factors that shaped the minds of our ancestors. And I no longer believe that. And the real key to changing my mind on that is I studied how people choose breeds of dogs for pets. And what I realized is that the role of culture was really much more important than I had realized."

Trends in dog choices are not necessarily good for dogs, as with the increased popularity of dogs with squashed faces, called brachycephalic features, such as French Bulldogs.[7] Brachycephalic dogs can suffer medical problems including respiratory, eye, and skin issues as a result of their looks. Several of these breeds (French Bulldogs, Bulldogs, and Pugs) are amongst the most popular breeds in the United States, Canada, and the United Kingdom (see table).

The most popular dog breeds in the USA, Canada, and the UK in 2018

	USA	CANADA	UK
1	Retriever (Labrador)	Labrador Retriever	French Bulldog
2	German Shepherd Dog	German Shepherd Dog	Retriever (Labrador)
3	Retriever (Golden)	Golden Retriever	Spaniel (Cocker)
4	French Bulldog	Poodle	Bulldog
5	Bulldog	French Bulldog	Spaniel (English Springer)
6	Beagle	Havanese	Pug
7	Poodle	Shetland Sheepdog	Retriever (Golden)
8	Rottweiler	Australian Shepherd	German Shepherd Dog
9	Pointer (German Short-Haired)	Bernese Mountain Dog	Dachshund (Miniature Smooth-haired)
10	Yorkshire Terrier	Portuguese Water Dog	Miniature Schnauzer

Source: Information from the American Kennel Club, Canadian Kennel Club, and Kennel Club.[8]

Dr. Jessica Hekman, a veterinarian who studies the role of genetics and the environment on dog personality at Darwin's Ark, recommends breed clubs support outcrossing projects. Instead of breeding dogs that are related to each other, outcrossing means mating two completely unrelated dogs (that is, ones that have no relations within a four-generation pedigree). This can bring in new variants of genes that will improve the health of the breed and help prevent problems due to inbreeding.

"WE CAN MAKE the world better for dogs by making dogs who fit into the world better. I would love to see dog owners draw a line in the sand and insist on dogs with muzzles long enough to let them breathe normally, or dogs who are not born with a 60 percent chance of developing cancer at some point in their lives due to their breed, or dogs whose heads are not too big for them to be born without a C-section. I'd love to see more breeders taking matters into their own hands and starting to experiment with how we breed dogs instead of continuing to use dogs from within breeds lacking in genetic diversity. I'd love to see more breed clubs supporting outcrossing projects to bring an influx of genetic diversity and healthy alleles into their breed. I'd love more dog lovers to become aware of the problems with how we breed dogs—how even the most responsible breeders breed dogs! This year, it is time for change."

—JESSICA HEKMAN, DVM, PhD, postdoctoral associate at the Karlsson Lab, MIT, and writer at *The Dog Zombie* blog

ANIMAL WELFARE AND OTHER CONSIDERATIONS

NO ONE WANTS to see Bulldogs, French Bulldogs, and Pugs disappear; they are lovely breeds with great personalities. But they should not have to suffer for their looks, and something needs to be done to improve the health of these breeds. Getting a dog is often a spur-of-the-moment decision and so people do not consider health. Research published in PLOS ONE looked at how the health of a breed plays into people's decisions to get a dog.[9] Four breeds took part in the study. Cairn Terriers were chosen because they are generally healthy, French Bulldogs and Chihuahuas because they tend to have health problems related to how they look, and Cavalier King Charles Spaniels because they tend to have health problems not related to how they look.

A survey of owners of these breeds found some of the problems were quite severe. For example, 29 percent of the French Bulldogs had had a sudden illness or injury in the previous year, and 33 percent of the Chihuahuas had had dental problems.

So why do people choose to get these kinds of dogs? Twelve percent of Cavalier King Charles Spaniel owners and 28 percent of Chihuahua owners said, "There wasn't really any planning" in the decision to get a dog. The personality of the dog, its appearance, breed attributes, and convenience were all factors given as part of people's decision. Owners of French Bulldogs, Chihuahuas, and Cavalier King Charles Spaniels often chose those breeds because of the dog's perceived cuteness, baby-like features, and fashion. In addition, people who were motivated by the distinctive appearance of the breed and by breed attributes were very attached to their dog. These results show people are not really taking the health of the breed into account. The scientists think this is because the emotional connection people have with a dog as a result of the looks can make a breed feel irresistible.

"CURRENTLY, DOGS ARE bred to meet human ideas of cuteness, with flattened noses, bulgy eyes, short legs ... Often these physical features cause ill health and suffering, such as the difficulties brachycephalic dogs have with breathing (who doesn't know a snorting, snuffling Bulldog?), with many other examples too. Meanwhile pedigree dogs are also, by definition, inbred from a small pool of animals of the same breed, and this directly causes an increased incidence in some illnesses (such as cancer in Flat Coated Retrievers). If humans stopped worrying so much about the cute appearance and 'breed pureness' of puppies, the resulting canine population would be healthier, and the dogs would be happier too."

—PETE WEDDERBURN, BVM&S, CertVR, MRCVS, veterinarian, newspaper columnist, and author of *Pet Subjects: Animal Tales from the Telegraph's Resident Vet*

Dr. Rowena Packer of the Royal Veterinary College in the UK has studied people's perceptions of brachycephalic breeds and the reasons why people choose these or other breeds of dog. In one study, she found that although many owners of brachycephalic breeds report their dogs wheezing, snorting, and snoring, more than half said their dog did not have breathing difficulties, which suggests people think these behaviors are "normal for the breed" (and also that some dogs are not getting needed veterinary treatment for these issues).[10] In subsequent research, she found that the kind of person who gets a brachycephalic breed is more likely to be a first-time dog owner than those who choose other breeds.[11] As well, those who chose a brachycephalic dog were more likely than those who picked other breeds to have used a puppy-selling website, and less likely to have seen the puppy's mom or asked about health checks.

"What came out of this analysis," said Packer, "is that owners of brachycephalic dogs were putting appearance as their number 1 influence as to why they were drawn to those animals. And I guess the concerning thing from a welfare point of view is that they were putting their dogs—that breed's health or that breed's longevity—as a lower priority."

"It was still a relatively low amount of people that said that they regretted their decision." What is interesting, said Packer, is that "A lot of people can't separate out the love of their individual animals from the behaviors that went into buying them. And I think we see that a lot with animals with chronic disease, that owners will say they would do it again and don't regret it despite the fact the animal is very poorly, because they love their pet." If they did have problems, they were likely to blame the breeder.

I asked Packer what people should do if they have a brachycephalic breed. She said, "If they already own one of those breeds, I think it would be a case of them being incredibly vigilant and making sure that they're not blinkering themselves with what issues those dogs might have. There's much better information out there online now in terms of what different health issues afflict different breeds." She also suggested taking any issues to a vet if you're not sure and getting independent veterinary advice. "The earlier a lot of them are diagnosed, the better the prognosis for any intervention," she said.

The bottom line is that anyone thinking of getting a particular breed of dog needs to specifically research whether there are health problems associated with that breed and if there are genetic tests the breeder should have done. The Institute of Canine Biology maintains a list of genetic databases on its website. As well, organizations such as the ASPCA describe the costs associated with owning different sizes of dogs (small, medium, or large).

Another thing to consider is whether it's the right time for you to be getting a dog. Do you have the time and energy to devote to a new dog? How do you think any existing pets will respond to a new arrival? Do you have any big lifestyle changes coming up that might affect your ability to care for a pet? For example, if you know you will be moving house, it's better to wait until you are settled so that a new dog won't have any extra disruption.

You should also think about the energy requirements of the dog, to find one who will match your lifestyle. And think about the dog's coat, because some will cover your clothes and furniture with dog hair and need lots of brushing and clipping, and others will require very little grooming. You may also need to think about allergies, which is tricky because you won't really know until the dog comes home. Better to get a pet you know with certainty no one is allergic to.

To make a good decision about a puppy, you also need to think about socialization.

ALL ABOUT PUPPIES: SOCIALIZATION IS KEY

THE PERIOD BETWEEN 3 and about 12 to 14 weeks is arguably the most important of a dog's life. This is the sensitive period for socialization, although there is some uncertainty as to the exact time when it ends (see table for the life stages of young dogs).[12] This is when puppies' brains are especially receptive to learning about the kind of social world they will live in as they get older. During this time they will also habituate to, or get used to, anything they might meet in later life (different sounds, surfaces, etc.). This is an important time in brain development, when the brain is very plastic and making lots and lots of new connections, some of which will be pruned out later.

The idea of a sensitive period sometimes surprises people, so it's useful to know other animals have sensitive periods too. In kittens, the sensitive period for socialization is between 2 and 7 weeks. This is typically before a kitten comes to live in your home, showing how important it is to get kittens from someone who will have socialized them. Children also have sensitive periods for development, during which important brain development occurs in response to the child's environment. These early life experiences provide the scaffolding for future development. If babies have lots of positive experiences with adults, very little stress, and good nutrition to help build a strong brain architecture, then by the time they start school they are in a better position to learn than if they had not had those experiences.

For dogs, lots of happy, positive experiences during the sensitive period for socialization mean a puppy is more likely to grow up to be a happy, friendly, confident dog. Bad experiences, or simply the lack of positive experiences, may lead to a fearful dog, as we know from classic research from the 1950s and 1960s.[13] So it's really important during this time that puppies are socialized to other safe dogs and to all kinds of people: men, women, children, older adults, people with beards and hats and backpacks and walking sticks, and so on . . .

The developmental stages of a young dog

PRENATAL PERIOD	Even though puppies are not yet born, there are already influences on their later behavior (see chapter 13 for discussion of the effects of stress hormones crossing the placenta). Prenatal learning can occur: puppies exposed to the scent of aniseed (added to the mother's food) can recognize it after birth.
NEONATAL STAGE: 0–2 WEEKS	Puppies are born with eyes and ears closed. Puppies cannot regulate their own temperature. The mother provides food and initiates elimination by licking. Puppies spend most of their time sleeping, and with their mother and littermates.
TRANSITIONAL STAGE: 2–3 WEEKS	The eyes and ears open. Puppies have a startle response. Early motor behavior begins as puppies start to move around. Early social behavior begins and puppies can wag their tails.
SENSITIVE PERIOD FOR SOCIALIZATION: 3–12 OR 14 WEEKS	There are many changes during this time as puppies start to learn all about the world around them. Puppies are weaned sometime between weeks 4 and 8 (varies by breed). Motor and social behaviors develop and become more adult-like. Puppies show more interest in people. Dogs learn social behavior through play with littermates. Vaccinations start at 6–8 weeks and continue until 16 weeks (with boosters as required).

JUVENILE PERIOD: 14 WEEKS UNTIL 6–12 MONTHS	Although the sensitive period is over, positive experiences during this time are still important as the brain continues to develop. Positive experiences will enable puppies to generalize from experiences during the socialization period. Puppies become more independent. A well-socialized puppy will want to interact with other people and animals. Growth continues (the timing at which it ends varies by breed, continuing longer for larger breeds). The juvenile period continues until puberty.
ADOLESCENT PERIOD: 6–12 MONTHS UNTIL 18–24 MONTHS	Post-puberty. Some female dogs can go into heat as early as 5–6 months.

Sources: Serpell et al. (2017), Bradshaw (2011)[14]

Puppies are typically adopted at 8 weeks old. Because those early weeks are important for socialization, you should research where your puppy has come from. Puppies that spent this period in a puppy mill or in a barn instead of a house, and then in a cage in a pet store, will have missed opportunities to be socialized (or potentially even had damaging experiences). Ideally, your puppy will come from a home (or a foster home, if a shelter puppy) where good socialization practices have been followed for the best possible start in life.

Pet stores and problem behaviors

Researchers use the puppy from a pet store as a proxy for the puppy coming from a commercial breeder. Studies show that puppies bought from pet stores are more likely to have behavior problems than those acquired directly from a responsible breeder. In one study, published in the *Journal of the American Veterinary Medical Association,* dogs from pet stores were more likely than dogs obtained from private breeders to be aggressive to their owner, to strangers, to other dogs, and to other dogs that live in the same house. Pet-store dogs were also more likely to have house-training issues and separation-related problems and to be sensitive to being touched.[15] Of course, there may also be differences between people who get their puppy from a pet store and people who obtain their puppy from a breeder (as well as in the information they are given).

A study published in the *Journal of Veterinary Behavior* took some of these owner-related factors into account and still found that 21 percent of the puppies from pet stores showed aggression towards their owner compared with 10 percent of those obtained from breeders.[16] Furthermore, this study found that puppies from pet stores had more issues with soiling the house, separation-related issues, and body licking; these particular problems were more common amongst owners who didn't attend dog training classes, took their dog only for short walks, and punished their dog on returning home. So there is an interplay between where the dog comes from and how the owner treats the dog.

While conditions in commercial breeding establishments vary, some are dire (and that's probably an understatement). Take a look at the ASPCA web pages on puppy mills to get an idea. Missing crucial socialization opportunities during the sensitive period is one reason for behavior problems later on, according to a report in the *Journal of Veterinary Behavior*.[17] Other reasons are genetic factors (e.g., breeding from animals that are already fearful of people); epigenetic changes due to stress (e.g., the mother being stressed during pregnancy); the puppy being weaned and separated from mom and the rest of the litter at too young an age; the stresses of being transported to the pet store, or being subjected to a restrictive home and/or pet store environment; and a lack of information being given to new owners on how to care for their puppy compared with if they had visited a breeder or shelter.

The only way to be sure your dog has been raised in a good environment is to see that environment. A study published in the *Veterinary Record* found that when people did not see either the puppy's mom or dad prior to getting a puppy, the dog was 3.8 times more likely to have been referred for a behavior problem as an adult; if only the mother was seen, the dog was still 2.5 times more likely to have been referred.[18]

If getting a puppy, make sure you see them with the mom (and be suspicious if the seller suggests meeting at a "convenient" location like a parking lot). Research any health checks recommended for that breed and ask about them. A good breeder—or a good shelter or rescue—will know about the importance of socialization and will be raising puppies in a home environment where they are becoming habituated to the sounds and activities in a normal household, so ask about socialization.

The power of more socialization

A study from the Guide Dogs for the Blind Association in the UK looked at the effects of puppies getting an extra socialization program on top of the existing (already excellent) one.[19] Six litters of puppies took part, all Labrador Retrievers, Golden Retrievers, or

Lab-Golden crosses. Between birth and 6 weeks of age, all of the puppies received the standard socialization program, and half of each litter also received the additional program. To make sure it wasn't just time with people that was making a difference, the puppies receiving only the standard program had someone sit with them for the same time as the new program took and interact with them if the puppies engaged with the person.

The new—additional—socialization program took just five minutes per puppy per day in the first week, rising to fifteen minutes per puppy per day in the fifth and sixth weeks. Taking account of what is known about the development of puppies and using resources that are readily available, the program included having a cell phone ring near the puppy; stroking the dog with fingers, with a towel, and with a hand in a rubber glove; and examining the puppy's ears and teeth. The new program had puppies experience these things on their own, away from the nest and their fellow puppies, as compared with the standard program in which puppies were socialized together. This may have helped the puppies become more resilient to separation-related issues.

The results were striking. At the end of the 6 weeks, there were already differences between the two groups of puppies. But the really important differences were seen in questionnaires completed by each puppy's handler when the dog was 8 months old. Puppies that had received the new program were less likely to have general anxiety, get distracted, have separation-related behaviors, or have body sensitivity (handling issues). These things matter a lot to guide-dog handlers, but they are also beneficial to all dog owners. These results show that extra socialization during those early weeks makes a difference to the behavior of the puppy as it grows up into a young dog.

The socialization period gives you crucial weeks at home in which to ensure your puppy has many different positive experiences. If your puppy is shy—and many are—take special care that they aren't overwhelmed. You can use play or food to try and turn situations into a positive experience (more on this in chapter 3). Give your puppy choices, encourage them to interact, but don't force it if they don't want to. Even if your puppy is shy, giving them a choice will let them come out on their own when they are ready—and that's exactly when the experience will be good for them.

Many humane societies and SPCAs have guidelines on how to choose a puppy, so it is worth looking online to find a guide from an organization you like. Some of them, like the RSPCA in the UK, also make a contract available that can be used when buying a puppy. Amongst other things, the contract will include something about what happens to the puppy if things don't work out; a good breeder or rescue will take the puppy back. Putting the time in to do your research now will help you find the right pet for you and ensure that your puppy had a good start in life.

ALL ABOUT RESCUES AND RETIREES

WHEN WE FIRST adopted Ghost, I was surprised that some people reacted very negatively to the fact he came from a shelter, even though right there in front of them was a stunningly beautiful— and well-behaved—dog. One man even told me that for sure Ghost would bite me. (Just so you know, he never did!) Aside from how rude these people were, they also were wrong: a shelter dog can be a good choice.

Research shows one of the things that often motivates people to choose a rescue dog is to save a life; not only do you give the dog you're adopting a new home, but you make space in the shelter or rescue for another at-risk or abandoned dog to wait for a new home.

I asked Dr. Sam Gaines of the RSPCA what advice she would give to people who are adopting a dog. She told me about the RSPCA's pre-adoption booklet, which includes information about what to do during the dog's first few weeks, including the importance of setting ground rules and having everyone in the house be consistent. At the same time, she says, people should pay attention to what they are told about the specific dog they are adopting.

Gaines said, "Put aside any preconceived ideas about the dog and how it looks, and I guess to a certain extent what breed it might be. Put that aside and instead think about who's the individual dog that you've just adopted." She added that it's important to work with the information you've been given. For example, "'This is what we've observed in that dog over the period of time that they've been with us; this is what their personality is like, things that they like to do' rather than assuming 'Well, he's a Labrador so he's bound to be really friendly and he's bound to love playing with a ball and he's going to be really safe with my family.' Moving away from that and thinking about 'this is his individual behavior and welfare needs and this is the species' behavior and welfare needs.'"

Be realistic about the dog you're getting

Just like when you get a puppy, it's a good idea to think about the commitment you are prepared to make to the dog. Dogs arrive at shelters for all kinds of reasons, such as when their owner has become sick, passed away, or been unable to find pet-friendly rental housing. But if the shelter tells you the dog has a behavior problem, it's important to consider whether you are the right person for that dog.

Dr. Carlo Siracusa is a veterinary behaviorist at the University of Pennsylvania School of Veterinary Medicine. As well as teaching vet students and seeing clients, he conducts research on the outcomes of dogs with behavior problems. He said, "Keep in mind that in many cases if you see that the dog looks like a very nervous dog—if he has a history, for example, of behavior problems, it's not that the dog was not trained; it's that that is the personality of the dog. So if you think you cannot deal with a dog like this or if it's not a dog that you want—like our clients say 'this is not the way we imagined the relationship with my dog'—then maybe you should not get that dog. There are people that are probably more experienced, that have already dealt with a more aggressive dog, with an anxious dog, with a dog with serious separation anxiety, and they feel that they can do it. Then that's okay. But do not think that because you love the dog that you will find a trainer that will do magic and will fix the problem."

A 2015 study of people who had adopted a shelter dog about four months prior found 96 percent said their new dog had adapted well or very well to their new home and 71 percent said the dog met their expectations: most of the dogs were friendly to visitors to the home and most never exhibited any of a list of problem behaviors.[20] Although 72 percent of respondents said there was a behavior they would like their dog to change—the most common being destructive behavior, fear, barking too much, and pulling on-leash—just over three-quarters of the people in this study said they would adopt from a shelter again.

Another study of people who adopted shelter dogs, published in *Applied Animal Behaviour Science*, reported that 65 percent were very satisfied with their new dog's behavior, and less than 4 percent were dissatisfied. On average, people rated their satisfaction with the dog as 4.8 out of 5. This study also reported that 53 percent of the dogs had a behavior problem, most commonly pulling on-leash, chewing or scratching furniture, or inappropriate toi-leting.[21] These results suggest that people realize they may need to teach their dog to fit into their household, and that many behavior issues are not particularly serious.

I don't know of a corresponding study that records how sat-isfied people are after bringing home a new puppy, but every dog needs to be taught how to be in your home. In the next chapter, we'll look at how dogs learn.

HOW TO APPLY THE SCIENCE AT HOME

- Consider whether you have the time, finances, and right home environment for a dog. Do you have time to exercise, groom, and play with them? If you will be out of the house for a long time on some days, can you make suitable arrange-ments, such as with a dog walker? Are you willing to learn about what the dog needs? You may like to babysit a friend's dog for a while to get some experience with taking care of a dog.

- Research the physical and behavioral health of the breed(s) you are interested in. Make a note of any genetic tests rec-ommended for the breed, and always ask about them. If a breed has a tendency towards health problems, you may want to choose the breeder carefully, take out insurance, budget for higher veterinary bills, or even choose a differ-ent breed. If you're getting a mixed breed to avoid hereditary issues, ensure the other breed(s) don't have the same issues (e.g., a cross of two brachycephalic breeds is still likely to be brachycephalic).

- Remember there are many breeds to choose from. If a particular breed's tendency to inherited health issues makes you think twice, reflect on what attracted you to that breed. If you wanted a small dog, research other small breeds. If you wanted a dog that doesn't need much exercise, consider appropriate breeds of all sizes (or maybe an older shelter dog would be right for you). If you are thinking of having children in a few years, pick a breed that is described as friendly—and make special efforts to ensure the puppy has nice experiences with children during the sensitive period (see chapter 8).

- Take advantage of the pre-purchase consultations offered by some veterinarians and dog trainers to help you think about the right dog for you.

- Always observe a puppy with its mom in their home environment before you take them home. Consider using a puppy contract (available from some animal welfare organizations).

- Ask the breeder (or foster home) how they are socializing the puppy. The sensitive period for puppy socialization is from 3 until 12 or 14 weeks. Make plans to continue socialization once the puppy is with you. A good puppy class may be part of your plans (see chapter 3).

- Give the puppy a choice, encourage (don't force) them, and protect a shy puppy. Remember that socialization means giving your dog happy, positive experiences.

- Don't forget that rescues and shelters are also good sources of family dogs; in some cases, an older dog may be a better match for your family than a puppy.

3

HOW DOGS LEARN

. . . .

WHETHER WE BRING home a puppy or an adult dog, we have to decide what the rules are and train them to behave the way we would like. Unfortunately, many people still believe the myth of dogs as members of a wolf pack, trying to be dominant all the time and competing with their human to lead the pack. It's a shame, because this idea immediately sets up the human-canine relationship as an adversarial one. But dogs are supposed to be our best friends. So let's start by looking at how dogs learn.

AN INTRODUCTION TO ANIMAL LEARNING

DOGS ARE LEARNING all the time, whether we are deliberately teaching them or not. Dogs learn throughout their lives, but they are also born with some species-specific behaviors called modal-action patterns (previously called fixed-action patterns, but the name has changed to reflect the fact they have some flexibility). Modal-action patterns have a genetic basis and are found in all members of the species, but they can be modified through learning. Hunting is an example, because some aspects of the hunting sequence are genetic, but they are also modified over time as dogs practice hunting skills. Other behaviors are entirely learned, either through interaction with the environment or through interaction with us. There are different ways in which dogs learn: non-associative learning and associative learning.[1]

Non-associative learning

Single-event learning is when a dog learns something after it happens only once. It can happen after eating something that makes them sick, just as with people when food poisoning or having too many of a certain alcoholic drink puts you off that food or drink in future.

Habituation is a simple type of learning that occurs when dogs gradually become used to something that is repetitive and not scary, so they no longer really pay much attention to it—like the background noise of the fridge or dishwasher. They are losing a behavioral response that was not learned, such as startling in response to the dishwasher, and they get used to the sound because they learn it doesn't mean anything to them. Sometimes they can dishabituate and pay attention to it again, but most likely after a short while they will realize it still doesn't mean anything and go back to ignoring it.

The opposite of habituation is sensitization, when an unlearned behavioral response (like a startle on hearing the dishwasher) gets worse and worse. If the dishwasher were dangerous, this would be a sensible response, as it would help the dog to avoid it, but since it is not dangerous it would just be an unnecessary source of stress. Sometimes people assume their dog will simply get used to something, and they accidentally sensitize the dog instead (see chapter 8 for how this can happen with children).

Sometimes it can be hard to know in advance if a dog is going to habituate or sensitize to a stimulus.

Social learning means learning from other dogs or from humans. Stimulus enhancement occurs when the dog's attention is drawn to something because another dog is manipulating it, and local enhancement means the dog is drawn to a stimulus or location because of the presence of another dog. Social facilitation means the dog will tend to do something, for example joining in running, because other dogs are also running. Dogs' abilities to imitate have also been investigated, particularly in terms of preferring certain foods, taking detours, or manipulating equipment to gain food.[2] Puppies that observe their mom doing narcotics detection work are quicker to pick up drug detection than those who don't, although it is not clear if this is due to observational learning.[3] The Do as I Do dog training method teaches dogs to copy a behavior performed by a human, insofar as canine anatomy allows.[4] However, more research is needed to fully understand social learning and whether there are simpler mechanisms underlying it.[5]

Associative learning

Dogs learn by association with events—that when the car turns in a particular direction it means they are going to the vet, for example. This is called classical conditioning and it affects the dog's emotions rather than their behavior. For example, if we know that a dog is afraid of strangers, we can make sure that the appearance of strangers predicts us giving the dog delicious food, and over time the dog will learn to like strangers.

Dogs also learn by consequences—if I jump up on you, I get to lick your face; if I sit when you ask, I get a peanut butter cookie. It's a simple concept (but it's so easy to accidentally reinforce the dog for doing something you didn't really want). This is called operant conditioning. When teaching dogs how to behave, we use operant conditioning to reward or punish behaviors.

Extinction happens when the dog learns that the consequences they were expecting no longer happen. Suppose every time your dog barks at the window, you ignore it. Your dog will continue to bark but—if nothing else is reinforcing the behavior—they will eventually stop barking. Before that happens, it's common to get something called an extinction burst, in which there is even more barking as the dog tries harder and harder to make the behavior work. At this point, people often think that ignoring isn't working and so they respond to the dog and inadvertently reward the behavior, which undermines the whole attempt at extinction.

Now I have to add a proviso, because dogs bark for many reasons and ignoring the behavior won't work if something else is reinforcing it (like the person they are barking at going away down the street). We can also accidentally extinguish behaviors we want by removing the reinforcement, like when we teach a dog to come when called using a food reward, and then just stop giving the rewards. The dog will keep coming back for a while, hoping for that cookie, but then they learn it isn't happening anymore. If other things are more motivating for them, they'll go and do those things instead.

Operant conditioning is the foundation of most dog training, while classical conditioning is often used to help fearful dogs. Let's look at these two types of conditioning in more detail.

Classical conditioning: learning from Pavlov

Most people are familiar with the story of Pavlov's dogs. Ivan Pavlov was a Russian physiologist who learned it was possible to pair a natural reflex like salivation with something totally unrelated (the sound of a bell). Dogs automatically salivate in response to the sight and smell of food, sometimes to the extent that drool dribbles out of their mouth. In technical terms, in classical conditioning we refer to the food as the unconditioned stimulus (US) and the salivation as the unconditioned response (UR). It is an unconditioned association because it happens naturally. Pavlov found that if he rang a bell just before delivering the food, the dogs would salivate in response to the sound of the bell. In this case, we refer to the bell ringing as the conditioned stimulus (CS) and the salivation as the conditioned response (CR). It's called conditioned because it has to be learned. It's not normal to salivate in response to the sound of a bell, but the dogs learned it meant food was coming.

Classical conditioning is most often used as counter-conditioning in conjunction with desensitization as a way of helping dogs to overcome fears. Desensitization means presenting the stimulus at a very low level that the dog is happy with, and gradually increasing it so the dog becomes used to it (the opposite of sensitization). In counter-conditioning, every single presentation of the stimulus is followed by something the dog likes (such as chicken or cheese) so the dog learns the stimulus predicts good stuff happening. Note that no behavior is required from the dog in desensitization and counter-conditioning (other than being aware of the stimulus), as the aim is to change the dog's emotions, not behavior.

Desensitization and counter-conditioning

- The "thing" (cs) happens at a level the dog is happy with—for example, a very quiet recording of fireworks or a stranger standing still in the distance.

- As soon as the dog notices the "thing," they receive food (us), which the dog likes (ur).

- Over time, the dog learns to like the "thing," which is the conditioned response (cr).

A great way to do this in real life is to use Jean Donaldson's Open Bar/Closed Bar technique. As soon as the dog notices the stimulus, start the flow of chicken or cheese (or whatever great treats you are using) as "the bar is open." Keep the flow going until the stimulus goes away or stops, and then stop the flow of treats ("the bar is now closed"). This technique helps to make the predictive relationship between the stimulus and the food obvious to the dog. All of this should happen while the dog is happy with the level of the stimulus. If you accidentally go "over threshold," immediately reduce the level of the stimulus (e.g., turn the volume down or put distance between you and the stranger), and then feed as per usual.

Operant conditioning: learning from Skinner

One of my favorite things to teach a dog is a brief sit-stay—especially if the dog is jumpy, bouncy, and mouthy, because it can make such a difference to the ease of interaction with that dog. It is fun too, with early steps that give the dog the chance to earn many rewards in a minute. Some dogs find it really tough to sit still while I dangle a piece of chicken in front of them for just one second; other dogs find the tough part is when I start to move a little and they want to jump up and follow me. Over time, as sitting still gets more and more of a reinforcement history, it happens more often, even when I haven't asked for it. This is in line with one of the early laws of animal behavior, stated by American psychologist Edward Thorndike as the law of effect: behaviors that get pleasant consequences will be repeated more often, whereas those that have unpleasant consequences will happen less.

B.F. Skinner elaborated on Thorndike's ideas and did the classic work on operant conditioning. He delineated what dog trainers often refer to as quadrants: positive reinforcement, negative reinforcement, positive punishment, and negative punishment.

Positive reinforcement (R+) means adding something immediately after a behavior occurs to increase the frequency of the behavior. Technically speaking, the term breaks down into two parts. Reinforcement means the behavior continues or becomes more frequent. And positive means something is added. For example, you ask the dog to sit, the dog sits, and you give them a treat (something is added). The dog is more likely to sit next time you ask (the behavior was reinforced). Here, the words positive and negative are not being used as evaluative terms (good and bad), but as neutral descriptions as to whether something was added or taken away.

Punishment means something that reduces the likelihood of a behavior happening again; in other words, the behavior becomes less frequent. So positive punishment (P+) means adding something after the dog does a behavior that decreases the frequency of the behavior. For example, if the dog jumps up when you come in the door and you knee them in the chest, and the next time you come in the door the dog does not jump up, you have positively punished the jumping. You added something (the unpleasant sensation of a knee in the chest) and reduced the frequency of the behavior. Please note, I am not advocating this as a way to train a dog, and we'll get to the reasons why in a moment. And it may also not work (e.g., if the dog perceives it as a game and keeps jumping). In everyday speech, when we say the word "punishment," we mean positive punishment.

Negative reinforcement (R–) means taking something away that increases the frequency of the behavior. An example would be pushing on the dog's bottom until they sit, at which point you let go. Assuming that the dog sits more often, the behavior of sitting is reinforced by removing the pressure on the dog's bottom. And negative punishment (P–) means taking something away that makes the behavior decrease in frequency. For example, your dog jumps on you and you turn away from them or even leave the room for thirty seconds every time. You are taking your attention away and the dog is less likely to jump up in future (but remember what I said earlier about extinction bursts!).

Examples of operant conditioning:
reward-based training uses R+ and P–

ANTECEDENT	BEHAVIOR	CONSEQUENCE	RESULT
You say "sit"	The dog sits	R+ Something good happens e.g., chicken, cheese, or treats; a quick game of tug; attention such as petting	The behavior happens more often
You come home	The dog jumps on you	P– Something good is taken away e.g., the flow of chicken, cheese, or treats stops; the game of tug ends; the person stops giving attention or leaves the room	The behavior happens less often
You greet the dog	The dog jumps on you	P+ Something bad happens e.g., a tug on the leash, pressure to push the dog's bottom down, a zap from the shock collar	The behavior happens less often
You say "sit" while tugging the leash, pushing the dog's rear end, or applying the shock collar	The dog sits	R– Something bad is taken away or stops e.g., tugging on the leash stops, pressure is no longer applied to the dog's rear end, the shock from the collar stops	The behavior happens more often

The table has examples of reinforcement and punishment. Note the consequence has to have an effect on behavior. For example, if you pet the dog intending it to be positive reinforcement but it has no effect on the dog's behavior, then the petting wasn't actually reinforcing to the dog.

Consequences are not the only way to change behaviors; we can also change the antecedents, something dog trainers call antecedent arrangements. For example, suppose the dog has a habit of drinking from the toilet bowl. The antecedent is that the lid is up allowing access to the toilet water. A very sensible antecedent arrangement would be to ensure the lid is never left up, so it is not possible for the dog to drink the toilet water. Of course, you should also ensure the dog has access to a suitable water supply!

DOG TRAINING: THE LINK BETWEEN TRAINING METHODS AND BEHAVIOR

REWARD-BASED METHODS ARE those that use positive reinforcement (R+) and/or negative punishment (P–), or humane management strategies (such as putting a lid on the garbage can to keep dogs from raiding the trash, or using a no-pull harness for dogs that pull on-leash).[6] Exercise and enrichment are also often part of the solution to resolving behavior problems (see chapters 9 and 10).

Eighty-eight percent of dog owners do at least some training at home, according to a report in the *Journal of Veterinary Behavior*, but it seems that most do not use reward-based methods exclusively.[7] Unfortunately, when people use outdated methods to train dogs, perhaps because they don't realize science recommends reward-based methods, they are using methods that rely on fear and pain. It's just a tap, a correction, or information, they say. But prong collars, choke collars, leash corrections, electronic collars, and alpha rolls (rolling the dog on their back and holding them there until they stop moving) work because they are painful or fearful for the dog. These are aversive methods.

A survey published in the *Journal of Veterinary Behavior* asked owners about their dog training methods and attendance at dog obedience classes.[8] The owners were then asked to look at a list of thirty-six possible dog behavior problems, including attention-seeking issues (e.g., jumping up, pawing, or mouthing the owner), fear issues (e.g., avoiding or hiding from familiar or unfamiliar people), and aggression, and indicate which one(s) their dog exhibited. Seventy-eight percent of dogs jumped up at people, 75 percent pawed at people or demanded attention, and 74 percent were excitable with visitors. These are all friendly, prosocial behaviors (at least in the eyes of the dog!). The three most common behavioral issues people described as problematic were aggression towards family members, house soiling when the owner was at home, and chewing or destroying things when the owner was out. Owners who used only positive reinforcement in training were less likely to report behavior problems related to fear, aggression, and attention seeking. Interestingly, the highest levels of fear, aggression, and attention seeking were found in dogs whose owners used both positive reinforcement and positive punishment (so-called "balanced" dog training methods).

A study in Vienna published in *Applied Animal Behaviour Science* looked at whether the size of the dog made any difference.[9] Dog owners in that city are required to register their dogs, and researchers sent a questionnaire to a random sample of owners, which means the results of the survey are representative of the population there. The study rated a dog as small (up to 20 kg or

44 pounds in weight) or large. Eighty percent of owners used punishment to train their dog, most commonly leash jerking, scolding, and holding the dog's muzzle. Ninety percent of owners used rewards either often or very often. For both small and large dogs, the more often their owners used punishment, the more aggressive and more excitable the dog. The relationship was strongest for small dogs. In contrast, the more often people used rewards, the more obedient they rated their dog, and also less aggressive and less excitable. Another finding of note is that owners of small dogs are less consistent with their training, put less emphasis on training, and engage in fewer activities with their dog than those who have large dogs. And consistency matters when it comes to obedience: the less consistent the owner, the less obedient their dog.

In another study published in *Applied Animal Behaviour Science* fifty-three dog owners were asked about how they had trained their dog and video was taken of them asking their dog to sit, lie down, and stay.[10] Researchers then gave the owner a bag of treats and a ball to use as rewards if they wished, and gave them five minutes to teach their dog a novel task—touching one of two spoons on command. All of the participants had used a mix of rewards and punishment to train their dog in the past. If the owners had tended to use punishment more often than rewards, the dogs were less playful with the owner and less interactive with the researcher. The dogs whose owners had previously used rewards more were quicker to learn the new task. Dogs also performed better at learning the new task if their owners were patient and used more rewards. The most likely reason for the improved results is motivation.

Another study, published in the *Journal of Veterinary Behavior*, observed dogs at two different dog training schools, one that used positive reinforcement and another that used negative reinforcement.[11] Dogs in the negative reinforcement group showed more signs of stress, such as a lowered body posture (keeping their body closer to the ground), whereas dogs in the positive reinforcement group looked at their owners a lot more. This matters because you need the dog's attention to ask them to do something. So positive reinforcement was not just better for the dog's welfare but also for the human–canine bond.

Gaze is an important part of the human–canine relationship. JEAN BALLARD

A questionnaire study published in *Applied Animal Behaviour Science* found that confrontational methods can lead to an aggressive response.[12] At least a quarter of dog owners reported getting an aggressive response to an alpha roll, dominance down (rolling the dog on its side and holding it there), muzzling the dog, forcibly removing something from its mouth, and grabbing the dog by the jowls. Use of a choke or prong collar got an aggressive response from 11 percent of dogs, and use of a shock collar got an aggressive response from 7 percent of dogs. Less aversive techniques such as growling at the dog, staring it down, or yelling "no" also sometimes got an aggressive response (yes, you read that right—some people growl at their dog).

A review in the *Journal of Veterinary Behavior* of seventeen journal papers about dog training methods, including those mentioned above, concluded that reward-based methods are better for dogs' welfare and in some cases even seem to be more effective.[13] Although many of these studies are correlational and so can't prove a causal relationship between training methods and signs of fear, anxiety, or stress, the existing research has led organizations such as the American Veterinary Society of Animal Behavior and the Pet Professional Guild to warn against the use of aversive methods in dog training.[14]

THE RISKS OF ELECTRONIC COLLARS

ALTHOUGH MANY TRAINERS get excellent results without them, some trainers still use electronic collars, also known as shock collars. Despite claims they merely "tap," "stimulate," or "tingle," they only work insofar as the dog finds the sensation unpleasant and worth avoiding. Otherwise these collars would have no effect at all (or the opposite effect to that intended). Alternately, some people say an electronic collar is a last resort, although the science does not support this view.

Research published in *PLOS ONE* shows that even when used by experienced trainers and in accordance with the manufacturer's guidelines, electronic collars pose a risk to animal welfare.[15] The researchers tested the collars specifically for training recall (dogs coming when called) in the presence of livestock (in this case, sheep). There were three groups of dogs: dogs trained with an electronic collar by a trainer recommended by the Electronic Collar Manufacturers Association, dogs trained with positive reinforcement by those same trainers, and dogs trained with positive reinforcement by trainers who specialize in using positive reinforcement. All of the dogs wore either an active or deactivated electronic collar so that observers who rated the videos could not tell which group the dogs were in (i.e., they were blind to the condition). The dogs in the activated electronic collar group more often showed signs of stress (such as low tail and yawns), although there were no differences in levels of the hormone cortisol (a measure of arousal). The study concluded that using electronic collars has risks for animal welfare and does not produce better results than positive reinforcement.

What about the use of electronic collars as a fence mechanism to keep a dog contained within a particular area? A fence is created by burying sensors underground along the line the dog is meant to stay inside, with visible markers above for training purposes. When the dog goes past one of the sensors, the collar delivers a shock.

A survey of dog owners in Ohio, published in the *Journal of the American Veterinary Medical Association*, found that 44 percent of people who used an electronic fence said the dog had escaped, compared with 23 percent of those who used a physical fence.[16] Unfortunately, if a dog escapes from an electronic fence (for example, to chase a passing cat), they may be reluctant to return to the yard because they will receive a shock on the way back in. Furthermore, these fences do not keep wildlife and other dogs or people out of the yard, meaning the dog is potentially still at risk of attacks from wildlife or dogs. Another risk is that the dog may associate the shock with the dog or person who just happened to be going past, and so may become fearful of or aggressive towards other dogs or people.

A review of the scientific research on electronic collars finds there is no justification to use them and suggests they should be banned; reward-based methods are encouraged instead.[17] Electronic collars (including fences) are banned in several countries, including Wales, Austria, Denmark, Sweden, and Switzerland (England announced a ban on other types of electronic collars but not electronic fences for dogs and cats).

THE BENEFITS OF REWARD-BASED TRAINING

ONE TIME, I spent an hour putting a harness on a dog and then taking it off again. She was a beautiful little Siberian Husky who pulled like crazy on walks and was not much used to being handled. Attempting to put the harness on resulted in excited jumping and mouthing, so I started by simply showing her the harness and giving her a piece of chicken for not jumping. Next I lured her head through the harness for chicken, then expected her to put her head through of her own volition, and so on. Our training progressed quickly, as she loved chicken, she was very clever, and she really wanted to go for her walk. Once I got the harness on, I took her out for a quick toilet break, and she was so happy to go outdoors. Then we went back inside for more practice at putting the harness on and taking it off again without her nibbling on my hands. Although I am used to wrangling jumpy dogs into harnesses, I wanted to know she would politely keep still while someone put her harness on. Siberian Huskies are known for needing lots of exercise, so this use of positive reinforcement enabled a future of long walks.

"WE ARE ULTIMATELY responsible for everything they experience, from their eating and elimination schedule, to their exercise and access to both wonderful and frightening things. Once we recognize that we humans are responsible for all of it, and that dogs are powerless animals whose welfare depends on us, kindness and consideration naturally follow. Dogs make choices when they have the opportunity—the choice to be warm, well fed, near the people and animals to whom they're attached (an important one!), and to be safe. We humans are the ones to present those opportunities.

Force-free behavior modification then makes sense: if you want to influence what a dog does, offer appropriate choices, give the dog time to choose, and reinforce the behavior you want. If the dog makes the wrong choice, try again—don't punish. Punishment leads to stress and unravels trust so that choice-making is inhibited. We are also capable of making choices; choosing to train dogs with kindness and generosity is an important one."

—ILANA REISNER, DVM, PhD, DACVB, Reisner Veterinary Behavior and Consulting Services

Training is good for your dog's welfare because it helps them know how to behave in order to get reinforcements such as petting, play, or food. In situations where a dog is unsure, they will default to behaviors that have been rewarded in the past, such as sitting. And, according to research in the *Journal of Veterinary Behavior* that looked at the outcomes for dogs referred to a veterinary behaviorist, good advice reduces euthanasias and keeps dogs in homes.[18] As well, reward-based training is a fun activity that can provide enrichment for your dog. At its best, training using food or play as rewards can teach dogs to detect narcotics or other substances, perform canine freestyle routines with their handlers, or even learn words, like Rico the Border Collie who learned over 200 words, and Chaser the Border Collie who knows over 1,000 words.[19]

"THE WORLD WOULD be a better place for dogs if every dog owner understood that their dog's behavior, good and bad, is motivated purely by consequences, not their dog's desire to be 'leader of the pack.' The myth that we must dominate dogs, or else they will assume the alpha position, is outdated and incorrect. Thanks to a recent explosion in the depth and breadth of canine research over the last 15 years, our understanding of dogs has improved dramatically. We now know that dogs are not trying to be the boss; they just do what works for them. Behaviors that have a desired consequence are repeated whereas behaviors that don't tend to stop. It's the same for us humans and, in fact, every other living being on the planet! This is why positive reinforcement training is so effective. When dogs (and other animals) are reinforced with things they like for desired behavior, they quickly learn to repeat those behaviors. Recent science has also taught us that physically punishing dogs (smacking; popping the check chain) for undesired behavior can adversely affect their welfare and the human-animal bond and punishment doesn't teach the dog what to do instead. Unfortunately, this relatively new understanding of dog behavior, learning and training has not become common knowledge amongst the general population and the old paradigm persists.

It's up to those of us who have this new understanding of dogs to share our knowledge far and wide to make the world a better place for dogs."

—KATE MORNEMENT, PhD, animal behaviorist at Pets Behaving Badly

Puppy classes

A good puppy class can help with socialization. A study in *Applied Animal Behaviour Science* found that attendance at puppy class was associated with a lower risk of dogs being aggressive towards unfamiliar people in the home or outside.[20] (This study also found that attendance at adult dog training classes was associated with a higher risk of canine aggression, perhaps because people are more likely to attend if they are having problems with their dog, or because of the methods used in the class, which were not assessed.) And we know that a one-off puppy party is not as effective as a six-week, reward-based puppy class.[21] Over a six-week class, puppies get the chance for ongoing socialization with other people and other puppies, and these encounters will help them to generalize those experiences. Puppies can start puppy class at 7 to 8 weeks of age and should have their first vaccines one week beforehand.[22]

A study by Dr. Janet Cutler, an animal behavior consultant at Landmark Behaviour and post-doc at the University of Guelph, asked new puppy owners what they were doing to socialize their puppy, whether or not they went to puppy class, and—if they were one of the 49 percent that did—what happened there.[23]

Cutler told me, "We found that people that did go to puppy classes were less likely to use punishment-based discipline, in particular yelling at their dogs or holding their dog down on its back. And we also found that the puppies of these people were less likely to respond in a fearful manner to some noises and also to crate training." The results are correlational, and it's possible that a different kind of person chooses to attend puppy class than those who don't.

The scientific literature does not have guidelines on how much socialization is enough, but for the purposes of Cutler's study, "not enough" was defined as up to ten new people and up to five new dogs in a two-week period. "The people that went to these puppy classes ended up exposing their puppies to more people and to more dogs," said Cutler. Still, about a third of puppies in this study were not getting enough socialization. She noted, however, that the quality of the experience is important. Forcing a puppy to meet people and dogs is not socialization; it is a potentially frightening experience that may do more harm than good.

Cutler noted that many classes don't habituate puppies to loud noises (such as fireworks), which may help them to not be afraid of loud noises as adult dogs. She also highlighted that many classes did not include handling exercises, which can help get the puppy used to the kind of handling they will experience at the vet's throughout their life. But she concluded, "I recommend that everyone takes their puppies to puppy classes, as long as it's one that's providing positive experiences. I'm a behavior consultant myself, and I have a puppy right now, and she's enrolled in a local school because I don't teach classes. So I'm going to puppy class with her even though I know about socialization, know what I should be doing. I still think that they're valuable things."

A good puppy class will ensure that all the puppies are having a good time by keeping shy puppies away from more boisterous ones, and letting puppies hide by their owners if they want to. Play opportunities must also be positive for all the puppies involved. If you're not sure, the trainer should do a consent test by separating the puppies. If the puppy who appeared to be victimized runs back to play, you know the play was okay. But if they don't, keep the puppies separate. A good trainer will ensure puppies don't get bullied and will use barriers or exercise pens as needed.

HOW TO CHOOSE A GOOD DOG TRAINER

TRAINING CLASSES ARE not just for puppies; adult dogs can also benefit from classes, including obedience or specialist classes on topics like greeting guests or liking visits to the vet. For behavior problems, private dog training is probably more appropriate.

When choosing a dog trainer, look for someone who will use food to train your dog, partly because this approach automatically avoids the use of physical punishment (such as prong collars, electronic collars, and alpha rolls) and because food is easy to deliver as a timely reinforcement. For some problems, it may be necessary to see your veterinarian, a veterinary behaviorist, or animal behaviorist (or a combination of your vet and a suitably qualified trainer). If no one is available locally, some trainers offer internet or telephone consultations.

"IF YOUR DOG is aggressive, scared, or destructive; or embarrassing, jumping up, and playing deaf... I promise it can almost certainly get better. Your dog can get better—and your relationship with your dog can get better. If you can commit to training your dog, however this ends up looking—taking a reactive rover class or working one-on-one with a qualified trainer, for example—there is help to be had. And when you come out on the other side with new skills for both human and canine, a slightly different setup at home, and some treats in your pocket, you'll be amazed that such a colossal change was even possible. So if things are tough and what you've tried isn't helping, reach out now. It gets better."

—KRISTI BENSON, CTC, dog trainer and staff member at the Academy for Dog Trainers

HOW TO APPLY THE SCIENCE AT HOME

- Use positive reinforcement, which is an effective way to train dogs and does not carry the risks of a punishment-based approach. If you want to know more about the science of dog training methods and how they affect welfare, I keep a list of research articles on my website with places where you can read about them online (look for the page called Dog Training Science Resources).

- To translate theory to practice, try to see problem behaviors from your dog's perspective. If your dog is doing something you don't like, remove the reinforcement for the problem behavior and/or provide better reinforcements for the behaviors you do want. Think about reward-based ways to manage the situation. If you think your dog is afraid, see chapter 13.

- Ask questions. When looking for dog training classes or hiring a private trainer, ask about the methods they use and ensure you are happy with the answers before you hire them.

- Look for a certified dog trainer who is a member of a professional organization, who takes part in ongoing professional development, and who will use food to train your dog. Well-respected certifications include the Certificate in Training and Counseling (CTC, from the Academy for Dog Trainers), the Karen Pryor Academy Certified Training Partner (KPA CTP) designation, the Victoria Stilwell Academy Dog Trainer (VSA CDT) designation, and the Pat Miller Certified Trainer (PMCT) designation. All these programs have websites where you can search for trainers.

- Look for a puppy class that emphasizes positive reinforcement and socialization, separates shy puppies from more boisterous ones during play, and encourages puppies (rather than forces them) to interact with other people.

- Set aside some time each day for training. A few short sessions are better than one long session. Make sure everyone in the house is on the same page, especially if you are dealing with behavior problems, so as not to undermine the training.

- Consider taking your adult dog to a class. Classes for adult dogs cover everything from basic obedience to fun activities like tricks or canine nose work (see chapter 10).

4

MOTIVATION AND TECHNIQUE

....

O NE TIME I saw a man walking a German Shepherd. Even from a distance it was clear the dog was nervous: his posture was low to the ground and the way he was walking made me wonder what kind of equipment was on him. As I waited at the traffic lights, I got a chance to see. The dog had a prong collar, tight, positioned high on his neck. There are easy alternatives, the simplest being a no-pull harness. I wondered why the man chose the prong.

We do know something about sources of training information. One survey published in the *Journal of Veterinary Behavior* found 55 percent of people said they got information about dog training from "myself."[1] This breaks down to 42 percent who got it from the internet, TV, or a book, and 13 percent who got it "instinctively." In another study, in *Applied Animal Behaviour Science*, "self" also rated highly as a source of information on particular techniques.[2]

Unfortunately, even when people turn to dog training books, they are not necessarily getting modern, science-based advice, according to research published in *Society and Animals*.[3] The study reviewed five dog training books chosen because of their ongoing popularity and was framed around what dog owners need to know. The silver lining is that some of the books contained very good information, and Victoria Stilwell's *It's Me or the Dog* and Karen Pryor's *Don't Shoot the Dog* both came out well. But the review found some popular dog training books include information that is inconsistent, scientifically inaccurate, or unclear; suggest the use of punishment-based methods despite their association with negative outcomes; and use anthropomorphisms and references to leadership that may interfere with dog owners' understanding of their pet's behavior.

Dr. Clare Browne, first author of the study, is a lecturer (equivalent to assistant professor) at the University of Waikato in New Zealand with a special interest in scent detection dogs for conservation. Browne told me in an email, "Good dog training books should have information that readers can understand and apply, but the information must also have a scientific basis. This review showed that not all of these popular books (that remained highly ranked on large retailers' websites for years) meet these functions. This is a concern, because people who read some of these books may not be getting the best information in terms of training efficacy and animal welfare." That is, this is bad news for animal welfare, and it's also bad news for owners who may struggle with their dog's behavior due to following poor advice.

LEARNING ABOUT DOG TRAINING TECHNIQUE

I'M TRYING TO teach Bodger to jump through my arms, and right now we've hit a tricky patch. I broke the activity down into steps and started by using food to lure him to step over my arm when I held it 10 cm (4 inches) off the ground. I soon realized that task was too hard and I had to start with my arm flat on the floor and my fingertips touching the wall. It's not the most comfy position for me, but now Bodger is very happy to follow the food and step over my arm. In no time I've dropped the food lure and he's following my hand; once all of him has stepped over my arm (and I get his tail in my face as he goes by), I get a treat from my back pocket to reward him.

The tricky part comes with the gradual raising of my arm. Five cm (2 inches) above the ground is fine because it does not disturb his natural step. When I move my arm a bit higher so it's at the level of his lower chest, Bodger's strategy is to try to barge his way through. "Too bad!" I say, trying hard to keep my arm in position. The same happens at the second and third attempt, only he tries even harder to force my arm out of the way. "Too bad!" I say. He knows this means he didn't earn his treat. And now it's time to lower my arm before the next attempt, because I'm following "push, drop, stick" rules as devised by renowned dog trainer Jean Donaldson. Four or five out of five correct and I push to the next level; three and we stick; but only one or two out of five and we drop back to the previous level. It's an efficient way to train because you proceed or drop back depending on how well the dog is doing, and this method helps to keep the dog's attention because that all-important rate of reinforcement stays nice and high.

We go this way for a few attempts—push, drop, push, drop—and then I split the difference to see if a smaller height increment will work. Bodger is totally focused the whole time, and it's fun because even when he is getting it wrong I move on swiftly to another attempt. When there is finally a little bit of a jump in his step as he goes over my arm, I'm delighted. And so is he, because his rate of reward has gone up.

This example shows all the important things about dog training: using a reward he likes (little peanut butter cookies), having a training plan that proceeds in gradual steps, dropping the food lure early on (but still rewarding with food every time), letting him know when he got it wrong ("Too bad!") and swiftly moving on to another trial, and getting the timing right so he is rewarded promptly. And, of course, he has a choice. If he gets tired and wants to stop, that's fine—the training plan will still be there another day.

WHAT MOTIVATES DOGS: FOOD, PRAISE, AND OTHER REWARDS

THE BASIC CONCEPT of training—the dog does the thing you ask and you reward them with food or a game of tug or something else they like—is nice and easy. But the detail is more difficult. And it's important to know how to motivate a dog if you want to train one.

A study by Dr. Erica Feuerbacher and Prof. Clive Wynne published in the *Journal of the Experimental Analysis of Behavior* involved five separate experiments, four with dogs and one with hand-reared wolves.[4] Two different types of dog were included: those who lived at home with their owners and those who lived in a shelter. It might be expected that shelter dogs, who received less human contact, would respond especially well to social interaction as a reward. But if a bonded relationship is needed for that interaction to be valuable, dogs with owners would be expected to respond more.

The animals were given the task of a simple nose-touch to the hand. In the food condition, dogs and wolves were rewarded with a small piece of food. In the social condition, the reward was four seconds of petting either side of the head and verbal praise from their owner or trainer (one of the wolves just received praise because he did not like physical touch). Both rewards took the same length of time. Across all three groups of animals—shelter dogs, owned dogs, and wolves—the food reward led to many more nose touches and a shorter time interval between them. Although there were individual differences, social interaction as a reward did not lead to many nose touches.

Another study by the same researchers, this time published in *Behavioural Processes*, investigated dogs' preferences for food versus petting.[5] Initially, two people—one offering food and the other offering petting—sat in a room, and as a dog approached one of them they received that person's reward for as long as they stayed within range. The researchers varied the setting (the dog's daycare versus a lab room at the university), the person offering the petting (the dog's owner or a stranger), and the preconditions to the experiment (briefly restricting, or not, the dog's access to their owner and food beforehand). They also tested shelter dogs. Over time the delivery of food changed from continuous at the beginning to a fixed schedule of one piece every fifteen seconds or one minute, to extinction (no food), and then back to frequent food delivery. The results showed that dogs prefer food over petting, and even though they made different choices when food was less available, they went back to the person delivering food when it was continuously available again. In a familiar environment but with a stranger offering petting, dogs preferred food, even when it wasn't very available.

But the results were slightly different for shelter dogs. Dr. Feuerbacher explained, "I sort of thought that they were going to choose food almost exclusively, and that's why we put in that thinning food schedule to see if we could push the dogs, under extinction, over to petting. And yet we found with the shelter dogs that as soon as food was slightly restricted, slightly less available, they opted for petting. And even some of the dogs preferred petting over food initially. That was surprising to me."

In comparison, food was much more important to owned dogs, and this only changed when the context was unfamiliar to the dog. Feuerbacher said, "One of the other effects we saw was that owned dogs would not alternate to petting very readily. They would wait even under extinction with the person that had given them food. And the only way to produce behavior that looked like shelter dogs was to have the petting person be their owner and put them in an unfamiliar environment. So we really kind of set up a situation where petting from Mom or Dad is much more desirable."

In another study, published in the *Journal of the Experimental Analysis of Behavior*, Feuerbacher and Wynne looked at dogs' preferences for petting or praise.[6] They gave 114 dogs (both shelter dogs and owned dogs) a choice between a person offering continuous praise (e.g., "Good girl!") and another offering continuous petting. After five minutes, the people swapped roles. Dogs preferred to stay near the person who was offering petting. Even when the person offering petting was a stranger, dogs with owners preferred to stay near that person. A follow-up study with different dogs found little preference between praise and no interaction at all. This research program shows that even though dogs like petting, the most effective reward is food. Other researchers have considered this too.

A study published in the *Journal of Veterinary Medical Science* looked at whether dogs' levels of interest in food might affect their ability to be trained.[7] Researchers took thirty-four dogs that lived in a kennel environment and divided them into three groups: those who did not finish a bowl of food and had leftovers, slow eaters who took their time to finish their food, and fast eaters who finished their food quickly. For three five-minute periods, they asked the dog to sit every five seconds, with a food reward every time the dog sat. Then, for another set of three five-minute periods, instead of using food, the handler said "good" and stroked the dog if they sat as requested.

Dogs in the fast-eaters group responded well to food as a reinforcement, and when the handler switched to praise and petting, the rate of response dropped off substantially. Dogs in the slow-eaters group responded well to food and also responded well to praise and petting. Finally, dogs in the leftovers group were not particularly motivated by either reward. However, *all* of the dogs responded better when food was used as a reward compared with praise and petting. So although there may be individual differences, all dogs are motivated by food.

Another study with fifteen dogs compared food with praise ("Good boy/girl") *or* petting rather than the combined praise *and* petting of the previous study.[8] The dogs were first given basic training in a sit-stay. Then they were divided into three groups according to the type of reinforcement used and trained to "come" using a standardized plan, with either food, praise, or petting as their reward. Although all the dogs needed roughly the same number of trials to learn to come, the dogs trained using food came significantly faster when called compared with those trained using either praise or petting. During the baseline training on sit-stay, dogs trained using food needed fewer trials to learn the task than those in the other two groups. The authors suggested that the type of reward is most important for the early training sessions, and that again food is the most effective way to positively reinforce desired behaviors.

What about the type of food? Good dog trainers will tell you to use good food, because although some dogs will work for kibble, many will not. In particular, use your best food for the most important behaviors such as coming when called. Recently, researchers tested nineteen pet dogs to see if quality or quantity of food makes a difference to running speed. The results, published in *Applied Animal Behaviour Science,* show dogs run faster when they know they will get a piece of sausage for their efforts compared with when they know they will get a piece of dry food.[9] But there was no difference in running speed when the reinforcement was five pieces of dry food versus one piece, suggesting quantity does not matter (the scientists did not test whether quantity makes a difference with pieces of sausage). Another study, published in *Scientific Reports,* investigated whether dogs prefer to get the same type of food or varied types of food as reinforcement.[10] These results showed individual differences, with some dogs responding better when the food items were varied, some when the food items remained the same, and some showing no preference. Over time, however, it seemed that preference for variety increased.

So it is up to you to find out what works best to motivate your dog and to vary it if needed to keep your dog interested.

"THE ONE THING I believe will make the world a better place for dogs is standards in dog training. The dog training business is unregulated which means anyone can take a course, hang a 'professional trainer' sign on their door, and give advice. This lack of accountability often results in information and training that is either ineffective, wrong and/or abusive. Every day I see dogs who suffer from training, handling and even owner expectations, along with owners who feel guilty because they're told to use techniques and tools that scare and hurt their dog. I truly believe owners don't want to hurt or scare their dogs and people get into the business of training to help dogs and have good intentions. It's the nonexistence of regulations that allow even the trainers to be misguided and misinformed. If we want to do better for dogs, demand training based on the science of learning theory and seek out trainers with the knowledge and skills to train both dogs and people using humane methods. If we bring dogs into our lives we owe it to them to do better by them. Until we have standards, remember we may not know what dogs are thinking, however we know they experience fear and anxiety, so have empathy when you're handling, caring for, and training a dog."

—**KIM MONTEITH,** CTC, manager, Animal Welfare, BC SPCA

Should we use a clicker in training?

A clicker is a small handheld device with a button that, when pushed, emits a short, distinct clicking sound. It is commonly used in reward-based training as secondary reinforcement, meaning that it tells a dog that something they find intrinsically reinforcing, like food or petting (primary positive reinforcement), is coming. In other words, this is a classical conditioning relationship in which the click means treat. The clicker (or if you prefer, a word such as "yes") marks the precise time when the dog performs the behavior that earns a reward. Many dog trainers use them, but are they more effective than if we don't use them?

A study in *Applied Animal Behaviour Science* divided fifty-one pet dogs into three groups and trained them on a novel task: how to open a plastic breadbox by pushing the handle up with their nose or muzzle.[11] Seventeen were trained using a clicker; another seventeen, using a marker word ("bravo"); and another seventeen, using a reward only. Using a method called shaping, in which the dog is rewarded for closer and closer approximations of what they have to do, the trainers taught the initial task and then tested the dogs with a similar task and a very different one to see how well they performed when asked to generalize the training. Almost all of the dogs completed both the simple and complex tasks. The scientists expected to find that using the clicker led to better results. In fact, there were no differences in the length of time taken or the number of trials needed for the dogs to complete the behaviors eight out of ten times.

A subsequent study by a different team and also published in *Applied Animal Behaviour Science* again put the clicker to the test.[12] This time people were invited to take part in a six-week, private, trick-training class in which an instructor came to their home to teach them how to train their dog to perform tricks. Again, the results showed no specific benefits or disadvantages to using a clicker in dog training in terms of the dog's impulsivity and problem-solving skills or the relationship between owner and dog. Essentially, both methods work, and people in both groups found the training fun and also challenging. There was a specific benefit to the clicker-plus-food over food-only for just one of the tricks taught, nose-targeting an object. The study also found, contrary to some dog trainers' beliefs, people were not put off by the use of a clicker.

Dr. Lynna Feng, first author of this study, told me in an email, "First, if you find that the clicker training is too difficult, for you or your dog or the two of you together, and you're just looking for a well-behaved pet, toss the clicker and just use food. Second, primarily interesting for those who teach puppy classes or general manners, even when first starting out with clicker training, the extra steps don't seem to discourage most people from training and having fun with their dogs. Finally, we found initial evidence that clickers helped owners feel that the training was less challenging for one of the more complex tricks."

It seems likely that reward markers such as the clicker are best suited to training in which precise timing is important. Research has also found that "clicker training" means different things to different dog trainers—some define it narrowly, and others, as any reward-based training.[13] More research on best practices in training would be welcome.

TRAINING TECHNIQUES: THE IMPORTANCE OF TIMING

HOWEVER GOOD YOU are at training, you can keep on getting better. The early steps involve coordination—getting the lure in just the right place to move the dog's nose where you want it or getting the treat inside the dog before they have done something else. Sometimes I feel like dogs experience time in a different way than me; I am slow and clunky compared with the lightning speed at which they jump up, sit down, sniff my jeans, and jump up again.

When teaching a dog to do a behavior (operant conditioning), the dog is learning that the behavior has a consequence. The consequence needs to be applied quickly to make this apparent (one of the reasons people sometimes use a secondary reinforcer such as a clicker). Dr. Clare Browne looked at the effects of timing on training.[14] She took video of people in dog training classes and analyzed 1,810 commands given to dogs. Forty-four percent of the time, she discovered, the dog did not respond. When the dog did respond, Browne looked at how fast owners were to praise and reinforce the dog. You won't be surprised to learn that some owners were very quick to praise the dog and very quick to follow that up with a treat. But some owners were very slow; the longest time was just over six seconds. This is a long time to a dog!

Browne taught three groups of pet dogs that a beep meant a piece of food would arrive via a feeding device. Then she trained them on a novel task: how to put their head into the correct one of two boxes. Infrared beams across the top of each box detected the instant the dog's nose went in, and a computer beeped and delivered the reward via a feeder. Some of the dogs got an immediate beep and then treat when they correctly completed the task (immediate reinforcement); a second group of dogs got both the beep and the treat one second after they completed the task (delayed reinforcement); and the third group got an immediate beep but had to wait one second for the treat (partially delayed reinforcement).

Browne told me about the results: "I found that with the immediate reinforcement, 60 percent of those dogs learned, so not all the dogs learned the task, but 60 percent of that group learned. But when I compared that to the one-second delay group, only 25 percent of those dogs learned. So there was quite a difference between those two." She explained that with the third group, the results were unexpected: "They came in at 40 percent learning, kind of between the other two, when they got the beep immediately and the food was delayed. You might have thought that they'd do better than that because they were getting some immediate feedback, straightaway they were getting that beep, which should have been functioning—we assume—as a conditioned reinforcer. But I suspect actually with the equipment that there was some other signal that was more salient to the dog, so before the feeder operated there was a slight click."

The results show the importance of timing in marking and rewarding when the dog does the behavior. Browne was interested in how this compares with real-life training. She noted, "I was watching people's body language because we know from heaps of other research that dogs are very receptive to communicative cues we give. Well I thought I was noticing that people were giving unintentional signals with their body movements prior to any intentional feedback. So before they were saying 'good dog,' I thought that they were already reaching towards their treat pouch or their pocket."

To test this theory, Browne set up another study in which she observed owners training their dogs to do simple tasks. "Was there a measurable movement made by them before intentional feedback? And I found yes, most of the time. In fact about 75 percent of the time the owners were giving quite a clear signal to the dog unintentionally, before they were giving intentional feedback. So 75 percent of the time they were making some distinct body movement, and most of those were hand movements. They were moving towards the treat pouch, for example. So I suspect that that was actually bridging that temporal gap," Browne said.

Her advice is to be consistent (e.g., always keep the treats in the same place so that your reaching movement is the same). Of course, professional dog trainers take pains not to have such "tells" when they are training. But the main takeaway is to be fast with reinforcement. "Don't muck around," she said. "Pay attention to your dog during training and try to give them feedback as soon as they have given you the response that you're looking for, as fast as you can."

TRAINING TECHNIQUES: THE IMPORTANCE OF PLAY

WHILE THE TIMING of rewards after a dog completes a behavior is important, what happens after training seems to make a difference too. Researchers who published in *Physiology & Behavior* set up two objects on separate pieces of cardboard.[15] They then trained Labrador Retrievers to distinguish between the two objects and choose the correct one by putting their front paws on the cardboard on which it was placed. If the dog completed the task correctly, the researcher clicked and gave them a piece of pork or chicken sausage. If the dog got it wrong, the researcher said "wrong" in a neutral tone. Once the dogs got the task right 80 percent of the time, the training session ended.

Then half of the dogs had a ten-minute walk to a place where they could fetch a ball or Frisbee or play tug, whichever they preferred, for ten minutes before walking back to the lab. The other half of the dogs were given a bed to rest on while the researcher chatted to the dog owner, but the researcher would call the dog's name to prevent them from sleeping.

The next day, all of the dogs returned to learn the same task again. And the result? Dogs who had taken part in the play session relearned how to discriminate between the objects much more quickly (twenty-six trials, on average, compared with forty-three trials for dogs who had rested). It is not known if these results are due to the hormones produced during the play session or because the play also included exercise.

In 2016, I interviewed Jean Donaldson to mark the twentieth anniversary of her book *The Culture Clash* and asked about the most common mistake people make when training a dog. Her reply? "It's not sufficiently addressing motivation. So, to put not-too-fine a point on it, basically failing to cough up the chicken." She has a very cute little dog called Brian, and I asked how she motivates him in training sessions. "He's very about primal nibs," she told me. "He's about this stuff called Rawbble, which is little kind of freeze-dried raw things. He'll work very nicely for chicken breast, and I cut it into tiny little dice. He'll work for cheese. He'll occasionally work for a toy but not much; he's not incredibly toy-driven and so I generally train him with food."

Food can also be used to help dogs learn to like something they are otherwise not very happy about—like trips to the vet, as we will see in the next chapter.

HOW TO APPLY THE SCIENCE AT HOME

- Find out what motivates your dog. Food is the best reinforcement for most everyday training situations, as all dogs like it and it's quick to deliver. You may need to prepare it in advance and have places where you keep it on hand to use as a reward (e.g., cookie jar at home, bait bag or pocket when on a walk).

- Experiment with different food items, and remember that variety can be a good thing.

- Use your very best rewards to train your dog on tasks that are important and/or difficult, such as coming when called when there are lots of competing motivations.

- Practice getting your timing right, both with using a secondary reinforcer (e.g., a clicker) if you're using one and with delivering the primary reinforcer (e.g., chicken). Deliver rewards fast.

- Use play and petting as rewards too, but note that praise on its own is not really effective.

5

THE VET AND GROOMING

·················

L AST SUMMER WE arrived early for Bodger's annual visit to the
vet so he could have a short walk down the street beforehand, as there are lots of interesting smells from all the other
dogs who have been there. When it was time for the consultation, the vet was friendly to him and—as always—made a point
of saying hello and patting him before getting started with the
exam. Bodger loves to be patted. He sat still while the stethoscope
was applied to his chest and cooperated while his ears and teeth
were examined, just like he's been trained to do. Any time Bodger
was good, I gave him some chicken that I had chopped ahead of
time and brought with me. I also used food any time he had something done that was potentially stressful. In fact, he didn't seem
to notice the vaccination, perhaps because of the chicken. On our
way out, the reception staff asked if they could give him a treat
too. Bodger seemed to really enjoy himself. It was a far cry from

the first times we took him to the vet, when he growled, snapped, and hated the experience from start to finish. It took a lot of work to get from that to such a happy visit!

From a dog's perspective, going to the vet is unlike anything else they do. As well as all the different smells from disinfectants and other animals (not just cats and dogs but maybe exotic pets too), there are bright lights, slippery floors, and unknown people. And then in the exam room, they may be prodded, restrained, and touched with pieces of equipment like the stethoscope. It's easy to understand why a visit to the vet can be a stressful experience—and if the dog finds it stressful, the owner may too.

One US survey found that although 85 percent of dogs had been to the vet in the previous year, a quarter of owners believe routine medical check-ups are not necessary.[1] Many people reported looking on the internet if a medical problem arose with their dog, some delaying going to the vet as a result. In some cases the dog may get more sick and thus require more costly and involved care once they arrive at the vet. While people reported concerns about the costs, 38 percent also said "my dog hates going to the vet" and 26 percent said "just thinking about going to the vet is stressful" (for the person).

Most people are aware dogs don't like to go to the vet, but there is a surprising mismatch between that knowledge and people's perceptions of their own dogs. In a study published in *Animal Welfare*, veterinarian and researcher Dr. Chiara Mariti had each owner and their dog sit for three minutes in a veterinarian's waiting room prior to an appointment.[2] The owner was asked whether their dog was stressed in the waiting room. A veterinary behaviorist watched a video of the waiting room wait and also provided an assessment.

Although both owners and veterinary behaviorists rated 29 percent of the dogs as highly stressed, they did not agree on whether a particular dog was highly stressed. Obvious signs of stress, such as the dog trying to hide or trying to leave the waiting room, were picked up on by owners. Other signs of high stress noticed by the veterinary behaviorists were the dog trembling, a low tail, lowered ears—and the dog trying to refuse to go to the examination room when it was time for the appointment. Of course, veterinary behaviorists are highly trained and so you would expect them to notice more signs than ordinary people, but learning to read a dog's body language is good for everyone. The most common signs of stress reported in the study were nose licking, panting, lowered ears, crying, grooming, and yawning, so these are all signs to watch out for in your dog.

Some dogs apparently know they are going to the vet even before they get there (of course this is especially easy if your dog gets in the car only to go to the vet). In a questionnaire study, Mariti found that 40 percent of owners said their dog knew they were going to the vet when they were in the car on the way there, and over three-quarters of owners said their dog showed signs of stress even before they got to the waiting room.[3] In fact, 6 percent of dogs had bitten their guardian at the vet at some point, and 11 percent had growled or snapped at the vet. Given that vet visits should be a regular occurrence in every dog's life, this is a problem. Also a problem is the fact that while half of owners said they could do some medical treatments at home, two-thirds said they sometimes found it difficult. And when they did find it difficult, 72 percent scolded their dog and did the treatment anyway. Mariti recommends not scolding the dog, trying to understand the source of the dog's stress, being gentle, and maybe asking for a behaviorist's help. She also recommends getting the dog used to visiting the vet clinic and to being handled.

This is, of course, easiest to teach when the dog is still a puppy. But if you're struggling to give a treatment at home, sometimes just going slowly and using food can help, or make an appointment for a vet tech to help you with the treatment, or work with a qualified dog trainer or behaviorist to teach the animal not to be scared of the treatment.

REDUCING STRESS AT THE VET

WHILE FEAR IS a normal emotional response to something frightening and can help an animal avoid a threat, anxiety is the anticipation of something bad (or something perceived to be bad). Generalized anxiety can be very bad for a dog's welfare. Because coercive and punitive methods, such as pinning dogs down or immobilizing them, can increase levels of fear and make it harder for dogs to be examined in future, the American Animal Hospital Association recommends the use of low-stress handling techniques.[4] The good news is that a growing number of veterinarians are trying to ensure low levels of stress for their patients whenever possible.

The late Dr. Sophia Yin pioneered the use of low-stress handling techniques for dogs and cats that aim to reduce stress in vet exams as much as possible. More recently, Dr. Marty Becker, author of *From Fearful to Fear Free: A Positive Program to Free Your Dog from Anxiety, Fears, and Phobias*, founded a certification program called Fear Free for veterinarians and vet techs, which "aims to eliminate things in the vet's office that bother dogs and cats—like white lab coats, harsh lights and slippery, cold exam tables—while adding things they like."[5] The program has expanded to include certifying veterinary practices, dog trainers, and other professionals, and it has a website for pet owners.

Becker told me he had an epiphany when listening to veterinary behaviorist Dr. Karen Overall give a talk at a conference in

2009. She compared veterinary care to children's experiences of the health care system of the 1950s and 1960s. Becker said she talked about "how fear was the worst thing a social species could experience and how it caused permanent damage to the brain. So those of us that are veterinary professionals are causing repeat severe psychological damage to pets by what we are doing and not doing. That behavior produces a physiological response, so behavior is medicine. And that we are not only harming them emotionally, we're harming them physically." So Fear Free is a revolution in veterinary care that aims to look after the emotional well-being of pets. Fear Free clinics use treats to help pets feel comfortable (and make a note of each pet's preferences for future visits). They change the lighting so dogs are not bothered by the hum of fluorescent bulbs (which their sensitive hearing can detect). They might use yoga mats so they can examine pets on the floor instead of on a table.

Becker said, "I think it's just the fact that pets have a broad range of emotions that we need to recognize... they have emotions and we have an obligation to look at both their physical and emotional well-being."

Do low-stress techniques make a difference? Scientists subjected eight dogs to two visits to the vet, one using traditional veterinary approaches and one using low-stress handling techniques.[6] The two visits took place seven weeks apart, and both included the use and removal of a muzzle, the use of a stethoscope, a basic examination, and a simulation of putting the dog into position to take blood samples and insert a catheter. The difference between the visits was that, in the low-stress exam, the dog had five minutes to explore the exam room at will both before and after the visit and was free to move around the room during the exam itself. The vet also tried to keep the dog's stress as low as possible and treats were available. The results showed that during the low-stress exam, dogs engaged in significantly less mouth licking, low tail, and whale eye, which are typically associated with fear and anxiety. While this is only a small sample, these are promising results that suggest low-stress handling is effective in improving the dog's perception of vet visits.

One of the things that helped Bodger get used to the vet was going there to sit in the waiting room for five minutes and be fed treats. Then we would leave. I asked the vet's permission in advance and went at times when it was quiet. This way, nothing scary happened; Bodger just had a nice time eating food and being made a fuss of. Once he was used to being in the vet's waiting room, we started to take him whenever Ghost had to go to the vet. Unfortunately, Ghost had to go often. But this was good for Bodger: whenever Ghost had an appointment, Bodger would go too and sit in on the consultation. And both dogs would be fed pieces of chicken or turkey or even cheese.

Using food

There are advantages and disadvantages to using food at the vet to help dogs (and cats) enjoy the experience more, according to one report.[7] If your dog has ever had to have surgery, you will probably have been advised not to feed them after 8 p.m. the night before. The main concern is that when the dog is under anesthetic for the surgery, the gastroesophageal reflex might cause the contents of their stomach to leak up into the trachea, which could cause aspiration pneumonia (a bacterial infection). However, most trips to the vet do not involve anesthesia, and so vets could be missing the benefits of using food during consultations.

Food can help the dog be less stressed, which in turn reduces the risk of the vet or owner getting bitten. If the animal is less stressed, then they are less likely to need sedation. Some people simply stop taking their dog to the vet because they find it too stressful, and using food to help the dog be less stressed could stop this from happening. And it's also a good opportunity for the vet to demonstrate how to use food in counter-conditioning dogs to scary things—a skill that many owners could put to good use if their dog is afraid of something else like fireworks, for example (see chapters 3 and 13).

Other benefits of low-stress veterinary visits are that the dog gets better detection of medical or behavioral problems, and the client has a higher opinion of the vet and their team and is more likely to follow advice or treatment plans.[8] The overall result? Better health and well-being for your dog.

Owners can also help reduce anxiety by comforting their dog at the vet.[9] Scientists had dogs visit the veterinarian twice: one visit in which the owner was instructed to sit quietly 3 meters (10 feet) away from the examination table and not interact with their dog, and another in which the owner was instructed to stand next to the exam table and comfort their dog. In this second group, owners spent a lot of time petting their dog and some time speaking to their dog.

In both conditions, dogs still showed signs of stress—such as lip licking and increased heart rate—during the exam compared with before it began. Overall, the dogs receiving petting and praise showed fewer signs of stress: their heart rate was lower, there were temperature differences (the surface of the eye as measured by infrared), and they made fewer attempts to jump off the exam table. But the researchers noted that sometimes owners touched their dog's nose or muzzle and/or held their dog by the collar, which many dogs do not like. These actions may have had the opposite effect than that intended.

SHOULD YOU SPAY/NEUTER YOUR DOG?

DOGS CAN BE prevented from breeding by spaying (removing the ovaries of female dogs) or neutering (castrating male dogs). Elective spay/neuter surgery is rare in some countries like Finland, but common in others such as the United States, where 83 percent of dogs are spayed or neutered.[10] Most animal shelters in the US, Canada, and the UK spay or neuter dogs prior to adoption to reduce the population of unwanted dogs.

The science on whether this surgery is good or bad for dogs' health is evolving. While some studies show an increase in longevity, others do not.[11] The surgery seems to decrease the risk of some conditions (such as mammary gland neoplasms, a common malignant tumor in dogs) and increase others (such as osteosarcoma, a type of malignant bone cancer, and overweight and obesity). Some breeds are at greater risk of these conditions than others, making it hard to know what is best for your particular dog from a health perspective. One very large study of dogs in the US and UK found there is a small longevity advantage for intact male dogs, but female dogs live longer if spayed (in fact, spayed female dogs live the longest of all).[12]

Spaying or neutering is sometimes suggested as a way to reduce aggressive incidents. One study of eleven breeds found no difference in how owners rated the trainability of their dogs, whether those dogs were spayed/neutered or intact (the exception: Shetland Sheepdogs and Rottweilers rated more trainable if neutered).[13] However, a study of over 6,000 male dogs that were neutered before 10 years of age found that longer exposure to sex hormones (i.e., later neutering) led to decreases in twenty-three of twenty-five behaviors mostly linked to fear and aggression (the two behaviors that increased were marking and howling).[14]

These results suggest sex hormones are important for behavioral development in male dogs, and neutering prior to puberty may increase unwanted behaviors. Ultimately, this means that while it's clear spaying and neutering leads to fewer unwanted dogs, the cost-benefit decision for individual dog owners is more complicated.

GROOMING

JUST AS DOGS have to go to the vet whether they wish it or not, they also have to be groomed. Dogs' fur varies so different dogs have very different grooming requirements. Bodger's fur is long and silky but curls up after getting wet in the rain, giving him a disheveled appearance and a kind of cowlick on his head. Ghost had very thick, soft, wolf-gray fur that he shed all summer long and that became even thicker in the winter months, with striped guard hairs on his shoulders and back. Somehow the mud never stuck to Ghost, but Bodger gets muddy with ease and twigs become coiled in his tail "feathers." Whatever kind of fur a dog has, we like to put our hands on it and feel it. And we have to do this in order to groom it.

To determine the response to grooming, scientists tested two very different groups of dogs—trainee guide dogs, who by definition were very used to being handled, and kenneled Greyhounds, who are kind of the opposite in that they are not normally petted.[15] The scientists picked four areas of the body (tail, saddle, chest, and ribs) and each part was groomed for eight minutes with a rubber grooming brush. In both groups of dogs, heart rate decreased during the grooming session, and there was no effect of the body part groomed (in other words, all parts were acceptable to the dogs).

Of course, this does not mean that all dogs like to be petted. In fact, although people commonly reach to pet the top of the head, most dogs don't like this; dogs generally prefer to be petted on the side of the chest or under the chin.[16] As well, it makes a difference if the person is familiar or unfamiliar, so always pay attention to body language and give the dog a choice to move away.

Similarly not all dogs like to be groomed and some positively hate it. The same goes for being touched in particular places, like the paws, or having their nails trimmed. This aversion may be due to a bad experience, but many dogs don't like it simply because as puppies they were never taught to like it. Indeed, puppies who have had the extra socialization program as part of the guide dogs study (see chapter 2) are less likely to be sensitive to body handling.[17] This is why it is important to do body-handling practice with puppies and to make sure their first-ever visit to the vet is a good one.

There's a name for when dogs are fearful or unhappy about being touched in situations such as bathing, grooming, nail clipping, and vet procedures: touch sensitivity. One study found that sensitivity to touch is more likely in puppies from pet stores.[18] It is also more common in shorter dogs and less frequent in bigger (taller) dogs.[19] It is important to remember that, in some cases, touch sensitivity may be due to arthritis and the dog being in pain.

In cases where people are unable to groom their dog or clip their dog's nails, it may be necessary to ask the vet for help. For your dog's well-being, do this before the fur becomes matted and their nails ingrown. Sedation may be required, and you can talk to your vet about whether this is a good option. Once the fur is shaved or brushed and the nails are trimmed, you can work out a plan to help the dog learn to like these activities in future. You may need to enlist the help of your vet and a dog trainer or behaviorist.

THE VALUE OF A HEALTH CARE COMMUNITY

PEOPLE WHO HAVE a stronger bond with their dog (such as having stronger feelings about the dog, spending more time together, and taking part in more shared activities) take them to the vet more often, with an average of 2.1 visits a year compared with 1.5 for those with a weaker bond.[20] Those with a stronger bond are also more likely to want preventative care (such as vaccinations and parasite prevention) and more likely to follow advice they are given by their vet. This carries through to the amount they say they would spend, if necessary, to save their dog's life. The bonding survey, done in 2008, found that people with a strong bond said they would spend an average of US$2,428 to save their dog's life compared with US$820 for those with a weaker bond with their dog. In fact, almost 20 percent of dog owners in this survey said they would spend "whatever it takes."

The American Animal Hospital Association (AAHA) recommends all dogs visit the vet at least once a year, but some dogs (especially seniors) will need to go more often (as will puppies for their vaccination schedule).[21] In addition to dogs' stress at the vet, other reasons for people not to take their dog for vet visits may be lack of awareness of the importance of these visits and concerns about cost. Costs and care can vary quite a bit between clinics. Veterinarian Dr. Adrian Walton of Dewdney Animal Hospital in Maple Ridge, BC, told me that differences in costs reflect differences in what is done to the pet; he called it "1988 medicine" versus "2018 medicine." Nowadays, he said, pets can get "basically the same quality of medicine as you would get if you went into surgery." He said higher costs typically reflect that quality, and, "if you're wondering why there is a different price between different vets, ask some questions, find out what the difference is. The vet should be able to explain it to you."

Studies by veterinarian Dr. Zoe Belshaw and colleagues at the University of Nottingham looked at owners' and veterinarians' perspectives on the annual veterinary appointment and discovered different expectations.[22] Owners felt they were not given much advance information as to what might happen at those appointments—information that is especially important for new owners—and those who had had several annual appointments said the visits tended to vary quite a bit. Owners thought a checklist might be helpful so they knew what to expect and that everything would be covered. Most vets said they had a checklist for the first puppy consultation, but they did not have one for the annual vaccination appointment for adult dogs. While owners thought the point of the consultation was really the vaccine, vets said this was only a minor part of it, suggesting that more information for owners would be helpful.

Newer dog owners tended to think the vet would use the appointment to vaccinate their dog and spot any health conditions their dog might have and therefore found the appointment "reassuring." In contrast, experienced owners did not think the vet would find any concerns they did not already know about. Vets, however, said they often find problems—most commonly lumps, obesity, tooth decay, and arthritis—that dog owners are not aware of. Belshaw said, "They've [the owner has] not done something awful by not noticing. That's why they need to go to the annual health checks, even if they're really confident that they've not noticed something. Because we do this all day every day, because we don't see the animal 24/7, we will probably find things that they've missed." Owners might not have felt the lump, paid attention to their dog's increasing girth, looked in their mouth, or noticed signs of pain, but all trusted that their veterinarian would tell them if they found something wrong.

Some vets in the study thought educating owners was a valuable part of the consultation, but others did not, either because they thought clients weren't interested or because they thought it wasn't a good use of the time available. And there were two topics that vets said they didn't particularly like to discuss: diet and behavior problems. Vets cited a lack of time, a lack of specific knowledge, and a belief that owners would bring up the subject if they were interested.

Belshaw's study points to the importance of the veterinary consultation in finding things the owner may have missed. Her advice for owners is to bring up topics you wish to discuss with your vet instead of waiting for the vet to mention them. If you have concerns, book a consultation instead of waiting for the annual appointment, and if you feel rushed during the annual consultation, book a single-issue consultation to discuss a particular issue in more detail.

The importance of the relationship between client and vet is another thing that came out of this research. Belshaw recommends that if you find a vet you like, you should stick with them. If not, it is worth shopping around to find someone you trust with opening hours and a location that works for you.

Once you have your team in place, she said, "Work side by side with vets. And also try to access any training that is available. So even if you think you're really experienced, stuff changes all the time, you know. If you have one dog at a time, and your last dog lived until it was fifteen, the stuff that you knew about puppies may already be fifteen years out of date." She added, "This is a really rapidly moving science. And so if there are classes that are available for you at the vet practice about puppy care or preventing disease or caring for an old cat . . . don't assume that there won't be something new for you to learn. And then don't be embarrassed about the fact that your vet might find something. It's a good thing; it doesn't mean that you're not caring for your animal very well. Engage with your vet and let them teach you more stuff because they're often very willing to do that."

DENTAL CARE

ONE OF THE most important and least discussed parts of dog care is their dental health. Teeth are important, not just because dental issues can be painful, but also because (just like in people) poor dental care can contribute to other conditions. One large study found a link between the severity of dental issues and the prevalence of subsequent cardiovascular disease, while another found a link with chronic kidney disease.[23]

The AAHA guidelines suggest that vets teach owners about providing good preventative dental care at home.[24] That care may include the use of oral rinses and gels, dental chews, dental diets, or additives to the water. The gold standard, however, is tooth brushing. One study found that in order to be effective, tooth brushing needs to occur at least three times a week, and daily if the dog already has gingivitis.[25] You can moisten and use a dog toothbrush, a child's toothbrush, or a finger toothbrush. Don't use human toothpaste, as it may contain additives that are harmful to dogs; special dog toothpaste is available with meat flavors to make it more palatable. Only try to brush the outside of the teeth.

When you have a puppy, it is much easier to make dental care a positive experience. Adult dogs may bite if they are not used to it, so find a knowledgeable dog trainer to help. If this kind of training is done properly (which includes slowly and carefully), it does not stress the dog at all. Remember that training doesn't just solve a problem, it can also provide enrichment. Another potential source of enrichment is friendships with other dogs, the topic of the next chapter.

HOW TO APPLY THE SCIENCE AT HOME

- If you have a puppy, get them used to going to the vet and practice handling them as if for a veterinary examination and for grooming, nail clipping, and tooth brushing when they are young.

- If your pet is already afraid of the vet, check with your vet about a quiet time to visit the waiting room. Bring food treats and toys, stay for five to ten minutes, and then leave. You may need to enlist a qualified dog trainer or behaviorist to help.

- Choose a veterinarian who uses low-stress handling techniques, and if you like your vet, build a relationship by returning to see them regularly.

- Check with the vet to see if you can help to comfort your dog during a physical exam by petting them in ways they like and using food or toys to make the visit a positive experience.

- Don't be embarrassed if your vet finds a medical concern during your dog's examination—after all, that's what they're trained for!

- Groom your dog more often for short periods of time. It's better to brush their coat, clip their nails, or brush their teeth for a short time and stop while they are still happy than to keep on going and make them anxious.

- Teach dogs—even adult ones—to enjoy routine grooming, nail clipping, and tooth brushing by teaching them to like the implement (shears and brush, nail clipper, or toothbrush) separately from teaching them to like it being used and any restraint that is necessary (such as holding the paws). You may need to ask a good dog trainer for help.

6

THE SOCIAL DOG

. . . .

WHEN HE WAS younger, Bodger had a great friend in a German Shepherd. The two of them were perfectly matched for size and play style, and they would become a rolling, growling, biting ball of fur until someone intervened to call them out of it. As soon as they were released to play again, the rolling, growling, and trying to bite at the other's neck would resume, one dog almost indistinguishable from the other. And that's one of the crucial things about play—both dogs were willing, consenting partners in the growl ball.

As they grew older, Bodger and his friend began to play less and just hang out more. This is a normal change for dogs, as from about 3 years old they play less overall and are more fussy about which dogs they will play with. But dogs are social creatures who extend their social circle to include us humans, as well as other animals we may have in the home (especially if they grow up with them).

And so dogs play not only with other dogs or with toys on their own, but also with us. But play between dogs can, to our eyes, look a lot like fighting, so how can we distinguish play from not-play?

WHAT DOES PLAY LOOK LIKE?

"PLAYTIME GENERALLY IS safe time," wrote Dr. Marc Bekoff, an internationally respected professor emeritus of animal behavior who's written extensively about play.[1] He noted that it is time in which animals cooperate with each other and are forgiving of accidental transgressions, especially when the play partner is young. It's in the interests of animals to play fair, he said, otherwise they might find they lose their play partners.

During unstructured play, dogs do a lot of biting at each other, and teeth can make contact with fur—but there are no yelps, and no injuries. That's because dogs are self-handicapping: they limit the force with which they bite, and also the force with which they slam into each other's bodies. For example, an older dog might not play as vigorously as they could when playing with a younger animal. Another characteristic of play is role reversals, in which dogs take it in turns to chase and to be chased, and to be the one on the bottom or the one on top when wrestling. An older dog might roll onto their back more when playing with a younger dog to make the play more balanced. Both self-handicapping and role reversals work to keep the play social and reciprocal, and to allow dogs of different ages, sizes, and abilities to play together. Also, the activities in play change often, from roughhousing on the ground one minute to running around the next.

Anyone familiar with dogs will recognize three play signals. One is called "play face," a lovely, happy, open-mouthed look that's a far cry from real fighting. Another is a wonderful exaggerated bouncy gait. And then there's the "play bow": front legs down, rear end up in the air.

The play bow

The play bow is a glorious signal but the reason for doing it isn't entirely clear. Traditionally, it was believed that the play bow signaled something like "I'm just playing, it's not real!" because many of the behaviors dogs perform in play—chasing, growling, biting, nipping—can also be aggressive. But if the play bow means "I'm only playing!" then you would expect to see more "offensive" behavior that could potentially be misinterpreted either just before or just after this signal. That's not what researchers found when they looked at play bows between adult dogs, in a study published in *Behavioural Processes*.[2] Instead, both the bower and the bowee were typically still before the play bow happened. Afterwards, play resumed in the form of chase sequences or both dogs rearing up. In other words, the play bow seemed to function as a signal to start play again after a pause.

The play bow. *JEAN BALLARD*

Dogs are not the only animals to play bow; other canids, such as wolves and foxes, do it too. In an experiment published in PLOS ONE, scientists looked at play bows in dog and wolf puppies.[3] The dogs and wolves in the study all grew up in a similar environment. That is, the wolf pups were born in captivity and hand-reared in small groups; the dog puppies were born in an animal shelter in Hungary and also hand-reared like the wolves. The researchers analyzed videos of dog-dog and wolf-wolf play in which at least one of the dogs or wolves was a puppy, and then coded the play bows that were performed by the puppies during a play bout.

It has been suggested that the play bow is a visual communication signal—that it is performed when the bower is in sight of the bowee—and this study and an earlier one in *Animal Cognition* found that to be the case.[4] Every one of the play bows done by wolf puppies was performed while the bower and bowee were in visual contact. And all but one of the play bows done by dog puppies was performed in sight of the other dog. In the one case where the other dog was not looking, the bowee barked, suggesting they knew they needed to get their partner's attention. As described for adult dogs above, if the play bow is a signal to say "I'm just playing," you would expect to see more "offensive" behavior immediately before or after it. Neither the wolf nor the dog puppies showed more "offensive" behavior than their partner prior to the play bow. However, unlike in play sessions involving the adult dogs, the puppies who were the bowee partner in the play sessions showed more offensive behaviors after the play bow, which is contrary to this hypothesis.

Earlier work by Bekoff found that play bows were associated with bite-shakes (biting and then shaking the head).[5] However, the scientists in this study found there were no bite-shakes immediately before or after the play bows. This is surprising, but the difference might be that Bekoff had looked at younger puppies.

In fact, there were few bites and nips in the videos of dogs and wolves used in this study.

It was also suggested that a play bow might position the bower either to run away from or to chase the other dog. In the earlier PLOS ONE study with adult dogs, there was no evidence it was being used to attack the other dog in play, but it seemed possible it was used to escape. In play sessions involving the dog puppies, the bowee was more likely to play-attack than the other way around. This result was not found in the play sessions involving wolves. Both wolf puppies and dog puppies were more likely to run away after the play bow, suggesting the bow positions them to escape. But wolf puppies did not have the brief moment of stillness prior to the play bow that was found in dog puppies. The function of the play bow for wolf puppies was not entirely clear. However, for dog puppies, it was shown that the play bow does not happen at random and does not appear to signal "I'm just playing." Instead, it serves to make play continue and often starts a new run away/ chase sequence.

"LET DOGS BE dogs. Let's appreciate them as individuals with unique personalities. Let them exercise their noses and all of their senses when they're home and out and about. Let them play with their friends and do zoomies to their heart's content. To appreciate what it's like to be a dog, we need to understand how they see, hear, touch, taste, and most of all, smell. We're most fortunate to have dogs in our lives, and we must work for the day when all dogs are most fortunate to have us in their lives. In the long run, we'll all be better for it."

—MARC BEKOFF, PhD, professor emeritus, University of Colorado, and author of *Canine Confidential: Why Dogs Do What They Do*

WHY UNSTRUCTURED PLAY IS
IMPORTANT FOR DOGS

SOCIAL DOGS CLEARLY enjoy play, but it must have evolved for a reason—or even several reasons, according to a review published in *Applied Animal Behaviour Science*.[6] One theory suggests that play teaches dogs how to control their bodies. In play, dogs fight, mount, chase, catch, and destroy—all involving motor skills they need to learn as part of their development. One of the things puppies learn from playing with their littermates is acquired bite inhibition—how they can bite without causing harm. The puppy that bites a fellow puppy too hard will find that play stops and will learn not to bite so hard the next time.

Another reason play is thought to have evolved is to develop social cohesion. From an evolutionary perspective, building social bonds with other dogs through play would lead to fewer fights, improved survival, and better success at reproduction. Dogs that have been properly socialized like to play with other dogs, and they enjoy play with their humans too. Play with toys and a person seems to be aimed at interacting with the person, and play can improve the dog–human relationship.

A third evolutionary reason for play is that it teaches dogs how to deal with unexpected events.[7] Changes in hormone levels in both the stress and reward systems and in the brain that happen during play are believed to help dogs learn how to cope with stressful events. Emotional overreactions or loss of control at other times could have serious consequences but are safer in play. Play itself is pleasurable and may be relaxing or exciting, so dogs can learn to deal with surprising events. Movements in play (including deliberate self-handicapping) help animals learn to recover, for example, from falls.

According to the review, there is evidence for all three of these reasons for dog play, and different stages of play appear to be important for different reasons. The beginning and end of a play session are especially important for social cohesion, while the main part of the play session is especially useful for developing motor skills and preparing for the unexpected.

This makes it sound like play is always good for welfare, but it's not. The same review noted that play can sometimes be a sign of bad welfare. For example, when play is a displacement activity because something stressful is happening, when individual play with toys reflects a poor environment or lack of attention from humans, or when a dog uses play to distract a human from punishing or hitting them. If play between two dogs is unmatched or one dog is bullying another, it can be bad for the welfare of the dog being bullied or hurt. And one study found that if people (in this case it was police officers) combine a game of tug with lots of commands and discipline, it can be stressful for the dog, as opposed to a tug game that is more spontaneous and affectionate.[8] So it's important to consider the context before deciding if play is good or not. Another threat to a dog's welfare may be if the dog is social—likes other dogs—but for whatever reason hasn't developed good play skills. It can be hard for dog owners to tell if dogs are really playing or fighting, and they may find it frightening or stressful to watch. A common response is to stop or limit the dog's opportunities to meet and play with other dogs—a shame for a dog who really wants to play.

Although we need more research on whether there is a sensitive period for the development of play skills, play with other littermates is an important part of puppies' learning. Once you bring your puppy home, opportunities to play with other puppies in puppy class are important for continuing to develop good doggy social skills.

Dog parks

In places where dogs are typically required to be kept on a leash, off-leash dog parks provide vital spaces for dogs to exercise and socialize. In one study at a fenced off-leash dog park in Canada, published in *Applied Animal Behaviour Science*, researchers looked at whether dogs find the experience stressful.[9] The scientists took saliva samples from eleven dogs before and after a walk, before a visit to the off-leash dog park, and after being at the park for about twenty minutes. The samples were analyzed for cortisol, which is a measure of arousal. While there was no difference between the cortisol levels before and after a walk, the levels were higher after twenty minutes at the park.

The researchers observed fifty-five dogs after they arrived at the park and found that in the first twenty minutes they spent 40 percent of the time with a human (sometimes with another dog too), about 30 percent of the time on their own, and about 25 percent with other dogs. Younger dogs were more playful whereas older dogs were more active. Eighty-three percent of the dogs displayed a play signal at some point, and most dogs showed a stress signal at least once. There was a correlation between play behavior and mounting, suggesting that to dogs mounting can be part of play. And generally speaking, owners' assessments of how friendly their dog was were accurate, in that the dogs rated most highly on friendliness on a questionnaire completed by their owners did exhibit more play behaviors.

The highest cortisol levels were found in dogs who visited the dog park least often, especially those observed to have a hunched posture, which means they may have found the experience stressful. In contrast, dogs who had already been to the dog park that week did not have as many stress-related behaviors.

In a second study, published in *Behavioural Processes*, which took place at the same dog park, researchers took video of what sixty-nine dogs did in the first 400 seconds (almost seven minutes) they were there.[10] Then the scientists coded the behavior they observed in a neutral way, describing what it actually was rather than applying pre-existing labels like dominant, submissive, play, or aggressive. In the first six minutes in the park, dogs spent about 50 percent of the time alone and about 40 percent with at least one other dog (the remainder was spent with humans or with a mix of dogs and humans). Over this period, the amount of time spent with other dogs decreased and the amount of time on their own increased. Younger and older dogs spent more time with other dogs than those who were in the middle of the age group.

Almost all of the dogs received and/or initiated contact from the snout either to the head or the rear end. This is not surprising because dogs use their noses to get information about other dogs. A dog just arriving at the park tended to make snout contact to the head of other dogs rather than to the anogenital region. Receiving this contact to the head was correlated with moving the rear end away from the other dog, which is in line with the idea that they are trying to get information about the other dog without themselves being sniffed.[11]

Scientists observed that if one dog chased another, that dog was not necessarily chased in return, suggesting the chasing behavior was not necessarily reciprocal, at least during the time period observed. However, starting to chase another dog did correlate with physical contact and with wrestling behavior. The researchers noted they saw hardly any behaviors that could be described as aggressive, not just amongst the specific dogs studied but from any of the other dogs in the park at the time.

These results are encouraging for owners who like to frequent the dog park; they also show that dogs spending some of their time there alone, rather than in full-on play all the time, is quite normal. Taken together, these studies suggest that if a dog appears to find off-leash parks stressful they may prefer not to go, but many dogs enjoy the experience, and the physical exercise and social opportunities with other dogs are probably good for them.

PET SIBLINGS: NEGOTIATING WITH OTHER ANIMALS

SOME DOGS CAN accept other animals, like cats, into their social circle. For safety's sake, it's essential to note this does not apply to all dogs, some of whom will only ever think of a cat or other small animal as food or as something to grab and shake. But some dogs and cats live in harmony and are even best friends.

A study in *Applied Animal Behaviour Science* asked people who have both a dog and a cat (or more than one of either) how they get along.[12] The researchers also visited the house to observe the dog and cat in the same room. The good news is that 66 percent of these cases showed the dog and cat behaved amicably towards each other. In another 25 percent of cases, they were indifferent. Unfortunately, in 10 percent of cases they did not get along at all. It was best when the cat joined the household first; the dog, coming into a house that already had a cat, seemed to adjust quite well to the existing feline. It was less likely to work out if the dog lived there first and then the cat came along. The age at which they were introduced was important too, as things worked better when a cat was less than 6 months old and a dog less than a year old.

The cats and dogs in the study often seemed to understand each other's communication, even though there were differences in the signals they used. For example, a wagging tail is a sign of friendship from a dog but of nervousness or impending aggression from a cat. But the cats and dogs seemed to be able to read each other's body language. The dogs had even learned a cat-friendly greeting. Cats often greet each other by sniffing noses, and the dogs in the study were observed to do this with cats. These nose-to-nose greetings occurred more frequently in the animals that had been introduced at a young age, suggesting early exposure to the other species enables it to learn their communication signals.

Another study, published in the *Journal of Veterinary Behavior*, also asked people with both a dog and a cat how well they get along.[13] Most of the time they were said to get on well, although the relationship was not close (e.g., they did not tend to groom each other). In this study, early introduction of the cat to the dog (preferably before the cat was a year old) was associated with a good relationship, whereas the age of the dog did not matter. The most important factor in the relationship was the extent to which the cat was comfortable with the dog. This could be because cats were domesticated less recently and so are not as good at sharing with other animals, but it is likely also because dogs can be a serious danger to cats. The relationship was also better if the dog was happy to share their bed with the cat, although cats were generally not willing to share their bed. Because it seems to be the cat's experience that mediates the relationship, if your cat and dog are not friends, you should put extra effort into helping your cat feel comfortable around your dog.

An early study raised four Chihuahua puppies, each with a mother cat and litter of kittens, from 25 days old until 16 weeks—that is, for the length of the socialization period.[14] After these early feline experiences, the Chihuahuas were presented with a mirror. They did not seem to recognize what they saw as a dog, in contrast to some Chihuahuas who were raised with other Chihuahuas;

these ones barked and responded to their reflection. After spending two weeks with other dogs, the Chihuahuas who'd been raised with cats did respond to their own reflection, showing they had learned what their species looked like. When the four cat-raised Chihuahuas were introduced to some other young puppies, instead of playing with the pups, they preferred to stay with their feline friends. This study showed that experiences with another species during the socialization period can make a difference to how a puppy gets on with that species. The implication is that if you want your puppy to become friends with a cat, they should spend time with cats during the sensitive period for socialization.

Companionship (being kept with or away from other dogs, according to the dog's needs) is one of the welfare needs, and play is a normal behavior. This means it is important to provide these opportunities in a way that your dog likes. For puppies, play opportunities with other puppies during the sensitive period for socialization can help them learn good canine social skills. Some adult dogs do not like other dogs, and if this is your dog then you should take this into account (don't take these dogs to the dog park!). Companionship from humans is also important for dogs, and this is the subject of the next chapter.

HOW TO APPLY THE SCIENCE AT HOME

- Make time to play with your dog. Whether you're playing with an object or directly with your dog, remember that real play is spontaneous and friendly; it may include praise and petting too. If you're giving lots of commands, it's a training session instead, and while it's great if training sessions are fun, there should be some play that is just play.

- If you get a puppy, make sure your puppy has playtime and socialization opportunities with other puppies to help them learn appropriate canine social skills. Safe play opportunities may be especially important for puppies from commercial breeding establishments (i.e., from pet stores and the internet) who may not have had an environment that allowed them to play much with the rest of their litter. The best way to provide these opportunities is via a puppy class that includes playtime.

- If your dog likes to play but has poor play skills, don't let them bully or irritate other dogs. Work with a good dog trainer to teach them better play skills. A great recall (coming when called) can help to call the dog out from play when they are becoming too much for the other dog (you may have to work really hard to get such a good recall). Bullies may need to be taken home from the play session promptly when they transgress, and then wait to try again another day.

- Recognize that as dogs reach social maturity, they play less with other dogs and they are choosy about the dogs they will play with. They are also less tolerant of dogs with poor play skills. Don't worry; this is a normal part of canine development.

- Be aware of the size of your dog's playmates. At dog parks, small dogs should be in a separate area from large dogs to prevent the risk of injury to small dogs.

- Tailor play sessions to your dog. If your dog likes going to the dog park, that's great. If they don't, that's okay too; some dogs find it stressful so don't keep taking them if they don't enjoy it.

- If you have other pets in your home, make sure each pet's needs are met and they do not have to compete with each other. Use pet gates to keep your dogs out of trouble by designating which parts of the house they have access to. If necessary, feed the animals in separate rooms or in dog crates.

- If you are planning to have a dog and a cat, get the cat first. If this isn't possible, try to socialize your dog around cats when they are a puppy. Remember, dogs and cats can become friends but it's not guaranteed.

7

DOGS AND THEIR PEOPLE

....

ALWAYS WATCH THE news after dinner, but once we'd had Ghost for a while, there was another reason to appreciate this habit. He took to lying on the other side of the room where he'd look at me. Before when I'd tried to make eye contact, his eyes would swivel away. But that changed. He'd watch me watching TV, and if I looked in his ice blue eyes he just looked right back. And his mouth was open in a lovely relaxed smile. I felt like he loved me. I'd say, "I love you!" and he'd give a little "woo!" back.

"I love you!"

"Woo!"

Surely he loved me.

Bodger alternates between nudging me or my husband with his head to demand to be petted. In his own insecure way, he loves us too.

Of course, Ghost was responding to the tone of voice rather than the meaning. When I said something he definitely understood—"Would you like to go for a walk?"—the response was far more excited. "Woo! woo woo woooooo!!!" He sounded like Chewbacca.

But do dogs really love us? What do we humans mean to them? There are several strands of research that look at this question, although scientifically it's not phrased in terms of love, which is a subjective experience. Instead, scientists have borrowed techniques from psychological studies with children and from neuroscience to investigate whether dogs have attachments to their humans.

Dogs have amazing social skills. They can follow a human's pointer to find the location of hidden food, whether it's pointing with the hand, nodding the head, gazing in the right direction, or putting a marker by the target.[1] As a result of clever experiments in which dogs are given the opportunity to "steal" food, we also know they understand a human's perspective (although not always).[2] They may have developed such impressive social skills as part of the domestication process, a case of convergent evolution. Some evidence for this comes from the Siberian fox experiment, in which foxes were bred for friendliness (see chapter 2). These foxes are as good as dogs at following human points and gazes, so it seems domestication has given us a special relationship with dogs.

DOGS AND ATTACHMENT TO HUMANS

ALTHOUGH ALL DOG owners know subjectively of their importance to their pet, there's a scientific perspective on this too. In psychology, attachment refers to a child's close emotional bond with their caregiver and was first described by John Bowlby.[3] Attachment is not just friendly behavior, because it includes wanting to be close to the caregiver in response to distressing events. In human children, there are four components to attachment: (1) distress when separated from the caregiver emerges at 7 to 9 months, (2) seeking proximity to the caregiver is especially apparent from 12 months when infants are suddenly more mobile as they are able to crawl and then walk, (3) a child's caregiver is a secure base from which they can explore, and (4) the caregiver is a safe haven to return to if the child encounters something distressing.

All four of these components can also be found in dogs with respect to their owners.[4] Dogs seek to be close to their owners, especially in stressful circumstances (proximity seeking). When the owner is not there, dogs can experience separation-related distress (see chapter 13). And experiments that borrow methods from child psychology have demonstrated both the safe-haven and secure-base effects.

A classic test of attachment between an infant and their caregiver, called the Strange Situation, is a standardized procedure in which the infant is in a room with their caregiver and then a stranger comes in. Following a strict protocol, the baby is left alone with the stranger, then is comforted by the caregiver, then is left all alone, and last is joined by the stranger again. An infant that is securely attached will be upset when the caregiver leaves the room but happy to see them return and easily soothed. Initial attempts to replicate the Strange Situation with dogs had mixed results, probably because a well-socialized dog is happy to see a friendly stranger. So scientists designed a version that included a threatening approach from a stranger.[5] Dogs experienced the threatening approach both with and without their owner present, and the order was counterbalanced (half with the owner first, half without the owner first). The analysis took account of whether dogs were reactive to separation (whined or barked while the owner was gone) or not.

The results showed some similarity to the behavior of infants during the Strange Situation test; the presence of the owner had a "secure base" effect. During the threatening approach, the increase in dogs' heart rate was not as great if the owner was with them. In the reactive dogs, if they first met the stranger when the owner was present, they were less stressed when subsequently meeting the stranger without the owner (though still not at baseline). The study also showed individual differences. Some dogs were interested in the stranger despite their threatening approach, whereas others reacted by growling and barking.

By 12 months, human infants will look to their caregiver to see their reaction in the event of something a bit scary, called social referencing, and this is found in dogs too. Social referencing has two parts: first, the infant looks from the slightly scary object to the caregiver, and second, the infant reacts (approach or avoidance) in a way that is influenced by the caregiver's response. As the slightly scary object that would make dogs feel cautious but not so scared they would run away, scientists in one study used an electric fan with green streamers attached.[6] Owners took their dog on-leash into a room with the fan, and as soon as the door was closed the fan was started by remote control. The owner released the dog and looked at the fan with a neutral expression. After a period of time, they responded with either positive or negative facial expressions and comments. Some dogs were confident to approach the fan. Of those who were not, 83 percent looked at their owner at least once after seeing the fan, and if their owner responded negatively, they were more likely to avoid it than if the owner gave a positive response.

One difference between this study and research conducted with infants is that the owner was neutral at the beginning. So the experiment was repeated and this time the person responded as soon as the dog looked over.[7] As well, both the owner and a stranger were present and one of them would respond to the dog ("the informant") while the other sat and read a book. Seventy-six percent of dogs with the owner as informant, and 60 percent with the stranger, looked to them for a response. When the informant was the owner, dogs in the positive group reached the fan more quickly and dogs in the negative group took longer to reach the fan compared with when the informant was a stranger. When the message was negative, dogs looked to the seated person more if they were the owner rather than a stranger, suggesting that dogs wanted information from their owner too.

These two studies show that dogs use gaze to look at a person for information when faced with something they are unsure about, and they explore or avoid depending on the person's response. The studies also show that the owner is more important than a stranger. With some small differences, these results are very similar to those found in studies of infants.

But do dogs always prefer their owners? Dr. Erica Feuerbacher and Prof. Clive Wynne tested this idea by giving dogs a free choice for ten minutes.[8] The dog's owner and a stranger sat down and would pet the dog if the dog came to them. The study took place both in the dog's home and in the unfamiliar setting of a university lab, which Feuerbacher said probably "looked and smelled sort of like a vet office" from the dog's point of view. About 80 percent of the time, the dogs chose to be near a person, but which person depended on the setting. In a familiar setting, the dogs spent twice as much time with the stranger, and in an unfamiliar setting they spent about four times as long with the owner. Even in the home setting, most of the dogs went to their owner first before they went to the stranger. (Interestingly, when shelter dogs were tested with two strangers, they typically showed a preference for one of the strangers, showing that dogs can establish preferences for one person over another quite quickly.) This study showed that while dogs are keen to meet other people, they have a special bond with their owner.

"In a familiar context the dog would sort of say hi, check in with the owner and then go and spend their time with the stranger," said Dr. Feuerbacher. "Whereas in an unfamiliar setting, the dog was really reluctant. I think some of our dogs never went and even said hi to the stranger, they just hung out with Mom in that probably stress-inducing context."

Feuerbacher and Wynne have tested dogs' preferences for petting, praise, and food (see chapter 4), so I asked what their research says about the relationship between a dog and their human. "We see a lot of owner effects," said Feuerbacher. "The only time we didn't see an owner effect was with the petting and vocal praise," referring to a study in which dogs chose petting from a stranger over praise from their owner. "It seemed that their preference for petting was so great that it overrode any preference for the owner. But in our other research we do see an effect of the owner, especially in these Strange or novel situations, you can see that the dog has a special interaction with the owner. And we did another research study looking at whether owner access was a reinforcer for the dog. I always joke with people that those of us that own dogs and can no longer go to the bathroom by ourselves already know the answer, that your dog will work to be with you. And so I think that research also points to the importance of the human–owner relationship to the dog. The dog wants to be with you. And our research suggests if you're not interacting with the dog, your presence to the dog is important."

There are important implications for how owners can help their dogs in a stressful situation. Feuerbacher suggested calm petting and being there for the dog would be beneficial.

CANINE NEUROSCIENCE AND DOGS' PREFERENCES FOR PEOPLE AND FOOD

ANOTHER STRAND OF research that speaks to the human–animal bond comes from canine neuroscience. Prof. Gregory Berns and his team at Emory University in Atlanta used positive reinforcement to train dogs to voluntarily go into and keep still in an fMRI scanner. This research looked at a part of the dog's brain called the ventral caudate, which, in line with earlier work on both monkeys and people, is activated in response to the anticipation of a reward.

One study looked at activation of the caudate when the dog was exposed to different smells: a familiar human (the dog's main caregiver), an unfamiliar human, a familiar dog, an unfamiliar dog, and the dog's own scent.[9] Twelve dogs took part, and the smells came from swabs taken from the armpit of humans and from the perineal-genital area of dogs. The caudate was activated significantly more in response to the smell of a familiar human than in response to any of the other smells, including the familiar dog. So dogs recognized the smell of their main caregiver and had a positive association with it. While it's hard to interpret these results in terms of the dog's subjective experience, it shows the importance of the person to their dog. In humans, the same part of the brain is activated in response to looking at photos of loved ones. It's unlikely that this is just a conditioned response to the caregiver being the one who feeds the dog.

Another study from Berns's lab investigated dogs' preferences for food, the sight of the handler who praised them, and a control condition in which nothing happened.[10] Since the dog has to keep absolutely still in the scanner, scientists did the experiment by pairing a different item (a toy car, a toy horse, or a hair brush) with each of the three conditions. Each item was on a stick and was presented to the dog for ten seconds, and then the relevant event happened. At a general level, there was no significant difference in activation of the caudate between the food condition and the praise-plus-sight-of-human condition, suggesting dogs found both rewarding. But there were individual differences between dogs: nine had roughly equal responses to food and praise, four preferred praise-plus-sight-of-human, and four preferred food only.

A further experiment in the same study violated the dogs' expectations by not following the toy car with the handler praising the dog. This experiment found that differences in caudate activation between getting praise and not getting praise were greatest for those dogs who had greater caudate activation in response to praise in the first experiment. This confirmed that the dogs did like praise. In a final experiment, dogs in a maze could choose to run towards a dish containing food or their owner who would pet and praise them. Most dogs sometimes chose food and sometimes chose the owner, but over the twenty trials they made different choices. The dogs' choices in this experiment correlated with the results from the first experiment. Of course, by definition these dogs are highly trained and the results may not generalize, but they suggest caudate activation is stable and predicts choices in an individual dog.

Berns told me in an email, "The takeaway is that dogs, like people, are individuals, and that there is a spectrum of motivations. Some prefer food, some prefer praise, and many like both equally. Know which your dog prefers!"

DO DOGS UNDERSTAND HUMAN EMOTIONS?

THERE'S EVIDENCE DOGS respond to human emotional expression too. In one study, pet dogs were tested in their home to see how they responded to their owner or a stranger pretending to cry or humming (a control condition).[11] Dogs paid significantly more attention when their owner was crying rather than humming. If the crying had made the dogs feel sad, you would expect them to go to their owners to be comforted. But the dogs directed their behavior to the person who was crying, regardless of who it was. Most of the dogs actually approached the person who was crying, and almost all did so in a way that could be seen as trying to comfort them. This doesn't necessarily mean the dogs had empathy, but they approached the crying person in a way that the person would interpret as trying to comfort them.

Another study had the dog's owner sitting behind a see-through door and either crying or humming.[12] In both conditions, dogs were equally likely to break through the door, but they were faster if the person was crying. Dogs who opened the door in response to crying had lower stress levels than those who did not; these dogs paced, fidgeted, and barked. As well, this study found dogs who were more attached to their owner were more likely to open the door in response to crying, which suggests the possibility they were feeling empathy.

As well as recognizing whether dog and human emotional expressions are positive or negative, dogs also recognize the emotion of the sounds that go with them (called cross-modal perception of emotion, since it includes sight and sound).[13] One study showed dogs pictures of dog or human faces that were either happy/playful or angry/aggressive. At the same time, the researchers either played sounds that matched or mismatched the emotion or played Brownian noise (a neutral, low-pitched sound like the roar of a waterfall). When the dogs were shown a dog face, the sound they heard was either a happy or an aggressive bark. When shown the human faces, the sound heard was a phrase spoken in either a happy or angry way in a language unknown to the dogs (Brazilian Portuguese). The dogs looked longer at the face when it matched the sound than when there was a mismatch. In contrast, when they heard the neutral sounds, dogs lost interest in the pictures. Dogs also paid more attention to the dog emotions than the human ones. This study was the first evidence that non-primates can recognize and match up the emotional expression from both faces and sounds.

HOW PEOPLE AND DOGS INTERACT

DOGS HAVE THEIR own ways of asking for things, like when they want tummy rubs. From infancy, humans use referential gestures to attract attention, such as pointing at toys they want. This ability has also been found in chimpanzees in the wild, great apes in captivity, and ravens, as well as in grouper fish and coral trout who use it to show other members of the species where prey is. Do dogs' gestures towards humans count as this type of communication? A study published in *Animal Cognition* analyzed 242 episodes of communication in videos of thirty-seven dogs interacting with their owners.[14] This was a citizen science project in which people took video at home and sent it in to the researchers.

The scientists found nineteen different gestures that, at least a lot of the time, satisfied the five criteria for referential gestures: (1) it is a gesture (rather than a movement that physically does something), (2) it is directed at an object or a part of their body, (3) it is aimed at someone else, (4) it gets a response from them, and (5) it is intentional in nature. Several things might show it is intentional, including waiting for a response and repeating the gesture if it has not worked to get what they want, or trying other gestures to get the outcome. In dogs, all five of those criteria were met.

The dogs' gestures were used to mean four different things: "scratch me," "give me food/drink," "open the door," and "get my toy/bone." Some gestures, such as pressing the nose or face against a person or object, could be used for all four meanings, while others were only used for some. For example, rolling over always meant "scratch me." The head turn, which involves looking from an object to the human and back, was the most common gesture. What is especially impressive about these results is that they apply to communication with another species (us)!

Play is an important part of interactions between people and their dogs.[15] Even when dogs have the opportunity to play with other dogs, they still like to play with people. When playing with humans, dogs are more likely to show toys and present them to their play partner, suggesting that play with people has a different motivation than play with another dog. Tug, chase, chasing objects, and showing objects all occur more often in play between a dog and a person than between two dogs.

People use a range of gestures to encourage dogs to play with them. One study looked at video of twenty-one people initiating play with their dogs at home without using toys, and followed it up by testing signals on twenty Labrador Retrievers.[16] In a surprising result, although some actions were better at encouraging dogs to play, they were not the actions used the most often. The most effective signals were when the human bobbed like a play bow, chased or ran away from the dog, or lunged at the dog. Tapping the chest or otherwise signaling the dog to jump up also often worked, but was not used very much. Some of the techniques people tried, including grabbing the dog by the scruff, stamping, and picking up the dog, never worked to induce play.

Another study found people showed positive emotions in just over 60 percent of play bouts.[17] This is typically the case in tug and tease (in which the person teases the dog, for example, by pretending to throw a ball or touching the dog's leg), whereas fetch was generally neutral from the owner's perspective. People smiled more when there was more direct contact, a closer proximity to the dog, and more movement by the person. Female owners had more physical contact with their dogs than male owners. This study suggests that only some activities are play for both human and dog. Perhaps the owners who provided the videos were basing their assessment of play on the dog's perspective.

The way people speak to dogs is very similar to the way we speak to human infants. For infants, a higher pitch, more variation in pitch, and a slower speed helps with language acquisition, but obviously dogs are not going to learn to speak the Queen's English (or any other human language). Research published in *Proceedings of the Royal Society B* found that people used pet-directed speech with dogs of all ages.[18] People speaking to puppies used a higher pitch, and puppies seemed to respond to this. When the scientists played examples of people reading a script in both normal speech and pet-directed speech, the puppies preferred the pet-directed speech but adult dogs had no preference. In another study in which dogs and puppies were played a recording of the same phrase spoken in either pet-directed or normal speech, they paid more attention to the pet-directed speech, as assessed by the amount of time spent looking at the speaker.[19] A subsequent study, which was more naturalistic because a person sat with the speaker so that the recorded voices were not disembodied, found that adult dogs preferred when people used both pet-directed speech and words relevant to dogs, such as "come here" and "who's a good boy/girl."[20] This finding suggests that talking to your dog in this way will help to build the relationship between you.

"I THINK THE world would be better for dogs if people stopped to consider things (all the things!) from the canine point of view. Dogs in companion and working roles are often put into situations they wouldn't choose for themselves. We can improve dogs' quality of life by considering our decisions that impact them—how long we leave them alone each day, where they live, what training techniques we use, how we transport them, what we expect them to tolerate (from interactions with children, to dress ups and involvement in other human pursuits—like sky diving) and ask ourselves at every step, 'Is this what my dog wants to do, if given a choice?' Not all situations where our dogs would choose differently are avoidable (e.g., temperature taking at the vet clinic) but people should consider dogs and the way their lives are lived from the canine perspective. To see dogs as individuals who can experience a range of emotions with the capacity to suffer or thrive, both physically and mentally, based on the decisions we make about their lives—rather than just assuming that dogs like what we like, or that they are there to meet our whims, provide us with utility benefit, or be our entertainment—would be a huge advance for many dogs in our world."

—MIA COBB, PhD candidate, Monash University, and co-author (with Julie Hecht) of the *Do You Believe in Dog?* blog

The research described in this chapter shows that dogs are pretty good at reading human emotions, using social referencing and looking to their owner when they see a new object, and using the owner's presence to help cope with a stressful event. The relationship between dog and human is one of attachment on the dog's part. The way the owner treats the dog, including whether or not they understand the dog's body language, may affect the relationship between dog and human.

We are responsible for providing everything our dogs need, and the way we do so can make a difference to how happy they are—or whether they're happy. And as we'll see in the next chapter, this is also important where children are involved.

HOW TO APPLY THE SCIENCE AT HOME

- Understand that you are important to your dog. Your presence can give your dog the confidence to explore new things, and your dog will look to you for information when presented with a new or stressful item.

- Know that your dog can tell whether you are happy or sad, and this may be why dogs seem to comfort people when they need it.

- Although there is no specific research on how long a dog can be left on its own, and it will depend on the dog, use four hours as your maximum guideline. If your dog has to be routinely left for longer than that, consider asking a friend or neighbor to pop in and see the dog, arrange for a dog walker, or find a good doggy daycare.

8

DOGS AND CHILDREN

....

THE SUNSHINE IS baking hot, but there's a chill in the air along the gravel path as the wind blows down the lake from the mountains. We are heading back from our walk past an observation tower and both dogs have big, happy smiles on their faces. Two children coming towards us are already asking their parents if they can ask to pet the dogs.

"Are they friendly?" asks the dad.

"Yes," I say.

"You can pet the smaller one," he says to his kids.

I can't say I blame him. Which would you pick: the one that is bigger than your smallest child and looks like a wolf or the medium-sized one that is cute?

"Ghost is the calm one," I say. "Bodger likes to jump up and give kisses."

I ask both dogs to sit. Bodger waits a moment while he is petted, and then licks the older child's face. She steps back, says

"Ugh!" and wipes her mouth, but she is smiling. Then they pet Ghost, just for a moment, and the littlest child delights in the feel of his fur in her hands.

"Thank you!" they say, running away down the path.

I like it when children ask before petting the dogs. Both Ghost and Bodger are great with kids and cope admirably when surprising things happen, like a toddler reaching out to touch them and promptly falling over. But while most people know you need to exercise caution around strange dogs and children, they let their guard down around dogs they know.

THE BENEFITS TO DOGS
OF INTERACTING WITH CHILDREN

HOUSEHOLDS WITH CHILDREN are the most likely to have a pet dog, and more than half of children with pets describe dogs as their favorite animal.[1] Many parents feel that a pet will provide company for their child and an opportunity to learn about responsibility through helping to take care of the pet (even though it's often the parent who really provides the care).[2] But what does it feel like to be in a household with children from the dog's point of view?

According to research by Dr. Sophie Hall of the University of Lincoln, UK, "The child–dog relationship has the potential to bring a number of benefits to the dog. For instance, having children in the home often imposes a strict routine, which transfers to the dog, resulting in regular and predictable meal times and walk times. Children also provide exercise and stimulation to the dog. Children often enjoy building obstacle courses for their dog, which depending upon the dog's age and health, can be a mutually enjoyable body and mind exercise! Of course, many children love their dog and, providing they show their affection gently and appropriately, can be a source of much happiness for the dog."

Hall and her colleagues interviewed the parents of children with neurodevelopmental disorders (autism and attention deficit hyperactivity disorder) and parents of children with typical development about the quality of life for the dog in their household.[3] The parents were also given a checklist to complete that included twenty-two behaviors related to stress in dogs, including lip licking, staring, blinking, walking or running away, whining, shaking, and hiding. One of the big advantages parents mentioned for their dogs was having a routine. Companionship with the child was seen as a nice benefit, and parents mentioned their dogs often approached the child to spend time with them. However, the parents noted that sometimes when their child initiated close contact, such as cuddling the dog, the animal saw it as stressful; the dogs generally preferred lower levels of contact such as petting or just sitting nearby. Another benefit for the dog was that the child provided opportunities for play, but parents also noted that it caused stress when the child was boisterous or accidentally ran a toy over the dog. Dogs seemed to enjoy coming close to parent and child or sitting with them during story reading.

Children's tantrums were seen as a source of stress for some dogs, although a few people said their dog did not pay attention to the outbursts and a few reported their dog chose to be near the child at these times, such as by lying on them (which service dogs are often trained to do, but these pet dogs just did naturally). Another source of stress was when their child hit the dog, and one parent had trained the dog to go away when this happened.

The study identified nine situations involving children in which parents may especially need to pay attention to dogs' comfort levels, ranging from meltdowns and tantrums, to having friends round to play, to playing with threatening toys (e.g., wheeled toys) near the dog. Parents said they managed their dog's stress in these situations in one of three ways. First, they provided a "safe haven," such as a quiet room, dog bed, or crate the dog could go to when things were stressful. Sometimes the dog would be sent to the safe place when the child began to have a tantrum. Second, parents noticed the dog saw them as a safe haven and would intervene when they realized the dog was getting anxious; if they weren't there, it was reported the dog would sometimes go looking for them. Third, parents taught their child how to interact with the dog, including to praise them, and how to train them. Parents of children with neurodevelopmental disorders especially talked about the importance (and difficulties) of teaching their child to safely interact with the dog.

Another study, with older children, around 10 years old, reported in *Anthrozoös*, found that being responsible for some of the dog's care—feeding, grooming, or taking the dog for a walk—was not linked to increased feelings of attachment to the dog, but it did increase the dog's likelihood of following the child's pointing gesture towards food.[4] It also was associated with more petting in a test of the dog's sociability with the child. In turn, when dogs were better at following the child's pointing gesture, the child seemed to have a stronger attachment to the pet. This suggests first of all that children notice how the dog responds to them, and secondly that there is an interrelationship between the dog's behavior and the child's attachment. Interestingly, dogs being petted less was associated with increased feelings of attachment, though it's not clear why this is so.

HOW TO RECOGNIZE WHEN A DOG IS
ANXIOUS AROUND CHILDREN

WE CAN TAKE it as read that a dog who bites is not happy in that moment (and nor is the recipient of the bite), but this behavior may also put the dog at risk of rehoming (difficult in the circumstances) or euthanasia. Children are at greater risk of dog bites than adults. According to the American Veterinary Medical Association, 359,223 children in the US were bitten by dogs between 2010 and 2012.[5] The risk is compounded by the child's size; their head and neck are relatively close to the teeth of a dog in any interaction, and consequently this is where they may get bitten. Young children are most often bitten at home by a dog they know, typically the family dog according to research in the US by Dr. Ilana Reisner and colleagues.[6]

Bites are often preceded by the child interacting with the dog, and a particularly risky scenario is the child approaching a dog who is sitting still or lying down. This means that close supervision of child–dog interaction is needed, and children should be taught to call the dog to them (and leave the dog alone if they choose not to approach) rather than to go up to the dog (which might disturb the dog). For older children, dog bites more often occur outside the home and involve a dog the child doesn't know. In many of these cases, the dog's owner is not present, and the bite would have been prevented if the dog had stayed in its yard. This shows the importance of letting people know if their dog routinely escapes from their yard and/or informing animal control as appropriate.

Unfortunately, people are not very good at recognizing risky interactions between dogs and children, and dog owners are even worse at it than people who do not have a dog, according to one study published in *Anthrozoös*.[7] People were asked to watch three videos of interactions between a young child and a medium or large dog. In one video, a baby crawls towards a Dalmatian who is lying down next to a ball; in another, a toddler walks around and touches a Doberman; and in the last one, a Boxer follows and licks the face of a crawling baby. All three interactions were risky, and the dogs were obviously showing anxious or fearful body language. But most people said the dogs were relaxed (68 percent) and confident (65 percent), and the dog owners were even more likely to say the dog was relaxed than the non-dog owners. Dog owners are more likely to assume a dog is friendly.

The study also found that people in general, especially those without children, tend to say "the dog is happy" or "the dog knows that it is just a small child" rather than pointing to particular aspects of body language. Every participant referred to tail wagging as a sign of positive emotions, which is worrying because only some tail wags are a sign of happiness (see chapter 1). Even when people did recognize the dog's emotional state as anxious or fearful, they were still likely to describe the interactions as playful or friendly.

Especially where children are concerned, we need to be aware that any dog can bite, and learn how to recognize signs of stress, anxiety, and fear in dogs. It's also important to know that young children are not very good at reading canine body language and will typically think a growling dog is "smiling" or "happy"—after all, they are showing their teeth.[8]

TEACHING CHILDREN TO INTERACT WITH DOGS

THERE IS A surprising lack of research into children's normal interactions with dogs. According to a survey in *Frontiers in Veterinary Science* of 402 people with a child under six and also a dog, the kinds of interactions children have with dogs change as they develop.[9] The study showed the need to take account of a child's development when supervising interactions with dogs. Children under 1 year old already have many interactions with the family dog, including petting the dog's head or body. In this study, dogs were most commonly reported as avoiding children between ages 6 months—when they suddenly become more mobile—and 3 years, perhaps because a mobile toddler can be scary for a dog. And though it was rare in this study for a child to be reported as doing something painful to the dog, when these incidents did happen, they were most often with children between 6 months and 2 years old. At this stage, children are still developing motor control, so may accidentally hurt a dog, and they are still developing empathy, so may not realize some actions will hurt. It is better to physically help young children pet the dog so they learn to be gentle.

All dogs should have a safe space where they can go to relax and will not be disturbed.
KRISTY FRANCIS

The study also found that older children were more involved in care activities, such as grooming the dog or being allowed to hold the leash, and were more likely to scold the dog or give commands. But while children of 2.5 to 6 years interacted more with the dog, there was a corresponding decline in parental supervision during these encounters. This is unfortunate because children in this age group may not realize their playful or caring behaviors can be frightening for the dog. Hugging or kissing the dog—both risky interactions—were more common in this age group. And while parents may expect their dogs to get used to the child over time, it is possible that the dog may instead become sensitized and more fearful.

A study published in the *Journal of Veterinary Behavior* revealed that most parents were not adequately aware of situations that would put their child at risk with a dog.[10] Parents were shown five photographs of child–dog interactions, four of which were situations where experts said the parent should intervene because they were risky for the child. These scenarios included a child crawling towards a dog that is resting on a blanket, a parent dangling a baby over a dog's head while the baby reaches towards the dog who is looking up, and a child holding the front paws of a dog that is lying on its back. Parents were asked how they would respond if the dog was familiar and if the dog was unfamiliar to the child. Results showed that if the dog was unfamiliar a majority of parents agreed they should intervene, but if it was a familiar dog, parents' guard was down and most said they would not intervene (whereas experts suggested they should). Furthermore, 52 percent of parents said they sometimes leave their dog and child unsupervised for a moment and 44 percent turn their back on interactions. Most agreed with the statement, "As long as the child is nice to the dog, he/she is allowed to play or cuddle up with the dog as much as he/she wants." Unfortunately, this is risky because most dogs don't like cuddles and may not perceive the situation as benign.

In better news from this study, very few people said they punished their dog for growling at the child, which is important because growling is a useful signal that the dog is not happy. Also good news is that when parents intervened, they did so in a friendly way rather than punishing the dog. It's a good idea to avoid punishment, because children may copy it and because studies show that punishment can sometimes elicit an aggressive response from the dog.

A review of studies that aimed to teach children from infancy to 18 years how to safely interact with dogs found that interventions had a moderate effect on children's behavior.[11] Some of the interventions involved showing videos to children or using computer programs, while others involved having a real dog for children to interact with. The effects of the interventions on children's behavior were larger than those on their knowledge. This is a surprising finding, because, in the field of health promotion, it's more common that people know they need to change their behavior but don't actually do it. For example, people might know that smoking is bad for them but not give it up.

Only some of the studies looked at behavior changes, and the researchers said these studies tended to be better designed and potentially also had higher-quality interventions. However, the reason these interventions may have had more of an effect on behavior than knowledge could be that it is easier to teach young children appropriate behavior around dogs rather than knowledge.

Taken together, these studies show that parents need to continue to supervise young children (especially as they become mobile), to intervene in situations with the family dog rather than assume it is safe because the dog is familiar, and to teach children how to interact with dogs.

PREPARING DOGS TO INTERACT WITH CHILDREN

THE MOST DIFFICULT situations for dogs with children seem to be when the dog is older than the child—that is, when the dog has already lived in the household before the child was born. Taking steps to prepare the dog for the arrival of a baby is very important.

One thing that helps dogs like being around children is having positive experiences during the sensitive period for socialization in puppies. A study published in the *Journal of Veterinary Medical Science* compared three groups of dogs: those who had been socialized around children during the sensitive period, those who had spent time around children but only after 12 weeks of age, and those who rarely had any contact with children.[12] To test the dogs' reaction to children, three 9-year-old girls were recruited to interact with the dog while the owner held the dog on-leash in a room. The leash prevented the dog from coming into any contact with the girls. In one situation, the girl entered the room, stood by the door, and called the dog's name. In the second situation, the girl approached the dog (stopping at a line drawn on the floor to keep her a safe distance away). In the third, she ran around the room for two minutes repeatedly calling the dog's name.

Dogs who had been socialized around children as puppies were not aggressive or excited in response to any of these scenarios. In fact, most showed friendly behaviors towards the child, although some showed escape behaviors such as looking away, especially in response to the running situation. In contrast, dogs who had only been exposed to children after the sensitive period for socialization were just as likely to be aggressive or excited as friendly. A few dogs who had never had much contact with children were friendly, but many more in this group showed aggressive behavior (defined as barking, growling, snarling, carrying their tail high, or wiggling their tail). This shows the importance of giving puppies nice experiences with children during the sensitive period.

"PROPER SOCIALIZATION AND habituation when young; this simply cannot be emphasized enough. Careful, positive exposure to all of the aspects of the human world they can expect to encounter as adults can greatly reduce their fearfulness when adults and help your dog to be emotionally stable and cope with the world it will live in. An important aspect of this is getting them used to vet clinics, by visiting regularly for simple weight checks and rewarding them whilst there, habituating them to traffic, and the sounds of parties and fireworks etc. Appropriate socialization with different types of people, children and dogs (again conducted carefully to ensure a positive experience) will also help your dog to cope with the social world and be better adjusted for its life-course."

—NAOMI HARVEY, PhD, research fellow, School of Veterinary Medicine and Science, University of Nottingham

Preparing a dog for the arrival of a new baby includes helping the dog make lots of positive associations with the baby. When the baby arrives, there will be changes to the dog's routine, new items associated with the baby, less attention given to the dog, and perhaps changes to where the dog is allowed in the house. Some of these are things you can get your dog used to beforehand, such as adding pet gates before the baby arrives and teaching the dog to walk nicely alongside a stroller. The more preparation you can do in advance, the easier it will be.

If you have a dog with behavior problems, having a young child in the house is not a risk factor for rehoming or euthanasia. Having adolescent children is more of a concern, according to research published in the *Journal of Veterinary Behavior* by veterinary behaviorist Dr. Carlo Siracusa and colleagues.[13] He explained, "[From] my experience in clinics actually, more than how busy the parents are with the kids is how much of a conflict there is between parents and kids. And in adolescence it's frequent: there's a lot of raising the voice, sometimes yelling, and in my experience sometimes the children can also use the dog as a proxy to become a conflict with the parents. They are told to not do something and they go and do this." He added, "Consider that our population of course is a special population; they are dogs that are presenting for behavior problems, not your average dog. But in general a dog with a behavior problem is a dog that usually shows a lot of anxiety . . . The dog that is anxious, that is reactive, will get more agitated when there is arousal in the house, when there are conflicts, when there is yelling."

It's hard to know whether this would apply to the vast majority of dogs, who have no need of a veterinary behaviorist. But it again shows the importance of giving dogs a safe space to go to if they are stressed. And teaching children of all ages how to behave around dogs is important too.

HOW TO APPLY THE SCIENCE AT HOME

- Supervise interactions between dogs and children very carefully; you should be close enough to intervene if necessary. At other times, use barriers such as pet gates to keep small children and dogs separate. As young children develop, they have better motor skills and are able to interact more with the dog, so don't reduce supervision.

- Teach children how to safely interact with dogs, and remember many interactions that look benign are actually risky—it's up to you to ensure safety. "Benign" interactions with a familiar dog are actually when most bites occur.

- Do not allow young children to approach a stationary dog (sitting or lying down), as this is a risky scenario for a bite.

- Ensure the dog has at least one, possibly several, safe spaces to go to if things get too much. These spaces should not be accessible to the child. For example, a crate that is made comfy with a nice dog bed in it, or a bed or sofa in a room the child cannot go into.

- Remember you are a "safe haven" for your dog. Be aware of the signs that they might be stressed (e.g., lip licking, looking away, blinking, moving away, hard eye, freezing, shaking) and be ready to help by ending an interaction, calling the child or dog away, giving the dog some nice food, or petting the dog.

- Make sure your dog has some quiet time during the day and somewhere quiet to be when things are especially noisy.

- Teach your dog how to behave around children; for example, not to jump up, which might knock a child over.

9

TIME FOR WALKIES!

BODGER HAS A good internal clock. He knows when it is time for his walk. He does not come to ask but keeps a close eye on me, often from the other room where he is also keeping a close eye on what is happening in the street. As soon as I get up, he jumps up and runs to get his squeaky lamb toy. While I am putting my boots on, he goes "*Squeak! Squeak! Squeak!*" with the toy. He runs with it still in his mouth to check if my husband is coming too. "*Squeak! Squeak!*" And sometimes, when I open the drawer to get his leash out, he drops the squeaky lamb in the drawer. While Ghost would want his walk in all weathers, Bodger is a fair-weather walker. In pouring rain, he will stick his nose sadly out of the door, his tail down, sniff the air, and then retreat inside.

On days when it is wet *and* windy, Bodger will let me put his harness on but will not even look outside the door, standing sadly, low tail swishing slowly, as if to say, "You surely don't want me to go out in that?!" Since the walk is for him, it's up to him. On those days I hope for better weather later in the day so he can still get out for a walk. But take today: beautiful sunshine, the creek and ditches running with snowmelt, people with and without dogs enjoying what feels like the first day of spring. The first downy woodpecker is already hammering away on a metal plate on the hydro pole, hoping to attract a mate, and the varied thrushes are tooting away like out-of-tune penny whistles. On such a day, it's glorious to be out with a dog, even if he insists on spending forever in the shade sniffing a mucky pile of snow.

I take Bodger for a walk twice a day, sometimes three times in summer when the evenings are light. There are many dog owners like me, but there are also many people who rarely, if ever, take their dog for a walk.

THE BENEFITS OF WALKING—FOR DOGS AND OWNERS

WALKS PROVIDE PHYSICAL exercise, help with weight management, and support dogs' welfare need for good health. By letting dogs sniff and do their own thing, they help fulfill the welfare need to engage in normal canine behaviors. By providing ongoing socialization and allowing interactions with other dogs and people (as appropriate for your dog), they also help support the need for companionship. In short, walks are good for both physical and emotional well-being.

Unfortunately, no research tells us exactly how many walks dogs need for good health, how much time they need to spend off-leash, and whether on-leash walks are enough to keep them trim. The American Animal Hospital Association recommends that exercise includes on-leash walks as well as games such as fetch or agility.[1] They also say to avoid extremes of hot and cold, and, for puppies and young dogs, to take into account the growth plates of bones (too much exercise may damage growing joints). For obese dogs, they recommend starting with a five-minute walk three times a day and building up to a daily total of thirty to forty-five minutes. To burn 230 calories at a brisk walking pace, a dog that weighs 45 kg (99 pounds) will need to walk 4.82 km (3 miles). As well, they point out that exercise provides ongoing socialization; helps the dog get used to different stimuli in the environment; and reduces stress, reactivity, anxiety, and owner-directed aggression.

It seems many dogs need more activity than their owners provide. Amongst the fifty most popular dog breeds in the US, only one is described by the American Kennel Club (AKC) as a couch potato: the Basset Hound. "Usually a daily walk at a moderate pace will fit the bill," the AKC says on their website, adding, "After a walk or play session, they'll typically settle down for a comfortable sleep." At the other end of the spectrum, the top-fifty list includes several breeds described as needing a lot of activity: Labrador Retrievers, Golden Retrievers, German Short-Haired Pointers, Doberman Pinschers, Brittanys, Weimaraners, and Border Collies. Of the Labrador, the AKC says "A Lab who doesn't get enough exercise is likely to engage in hyperactive and/or destructive behavior to release pent-up energy." And of the Golden, "A Golden who doesn't get enough exercise is likely to engage in undesirable behavior." How much exercise is enough for a Lab or Golden?

A study of Labrador Retrievers in the UK, published in *Preventive Veterinary Medicine*, found the average Lab gets 129 minutes of exercise a day, most of it off-leash or in unspecified activities (which don't include on-leash walks/runs, playing fetch, or obedience, but possibly include work).[2] Working dogs got more exercise than pet dogs, and those living in homes with children got less exercise. People who said they had to restrict their dog's exercise due to the environment in which they lived were more likely to say they had a dog who was overweight or obese.

Another study of 1,978 Labs in the UK, published in *Applied Animal Behaviour Science*, found the dogs got anywhere from less than an hour to more than four hours of exercise per day.[3] Dogs who got less than an hour of exercise per day were more likely to get agitated if ignored, to bark, to show fear of humans and objects, and to be excitable compared with dogs who were exercised for more than four hours a day. Other results showed that dogs who got less exercise were more likely to show aggression to their owner or to other people and to show unusual behaviors. Since dogs who got more than an hour of exercise per day were rated as more trainable, we have to wonder if some dogs are walked less because it's difficult for the owner in some way.

Overall, these results suggest that without enough exercise, Labrador Retrievers may become bored and frustrated and may develop behavior problems, and this likely applies to other breeds too. Working dogs are bred to work for long hours and need more exercise, so dogs bred for conformation may be better as pets.

An Australian study found that 36 percent of people exercise their dog every day, 28 percent more than once a day, and 8 percent once a week or less.[4] Seventy-three percent said they took their dog for on-leash walks, 50 percent played games, 36 percent had off-leash walks, and 61 percent let the dog exercise alone in the yard or house. For pets with mobility issues for whom walking is difficult, swimming can be a good option, and mobility aids (such as a harness with a handle for the owner to support the back end of the dog) and physical therapy may also help.

THE HUMAN CONNECTION IN DOG WALKING

SINCE IT IS no longer permitted or safe for pet dogs to roam as they will in most places, dogs rely on their people to take them out for walks. Few of us meet the recommended minimum amount of exercise needed for our cardiovascular health. To state the obvious, walking the dog more would be good for us and good for the dog. Because of the contribution it could make to human health, scientists are very interested in what motivates people to walk their dog.

One of them is Dr. Carri Westgarth, a researcher at the University of Liverpool, UK, who studies dog walking and how best to prevent dog bites. In one study, she observed what people and dogs do on walks at three different dog-walking locations in the northwest of England—a beach, a sports field, and a field surrounded by woods.[5] She found that most dogs arrived with one person (59%), and most dogs were off-leash (73% during the week, 59% at the weekend). When the dogs were off-leash, sniffing was much more common compared with when they were on-leash. Not surprisingly, sniffing was associated with peeing and pooping, as dogs often sniff before toileting. The dogs' social interactions were more often with other dogs than with people, but there were more interactions with other people if the dog was off-leash. One of the implications is that if your dog is on-leash, you should allow more sniffing time.

I asked Westgarth why it's important to take dogs for walks. "My view is it's important for the dog mainly," she said, "not just physical exercise to tire them out and keep them fit but mental stimulation as well. But I also think it's important for the owner because it's one of the main joys we get. From my research it seems to be one of the main joys about having a dog is seeing it run around on walks and spending that time with it, and getting the stress relief ourselves from that experience. Assuming that the dog walking is easy and pleasurable."

Westgarth found dog walkers are not really motivated by the idea of their own better health but more by their dog's happiness.[6] She thinks we can get more people to walk their dog by letting them know that it makes their dog happy. In Westgarth's study, people's accounts of dog walking were framed around what was right for the dog and the responsibilities of being a dog owner. Some people talked about dogs that really loved to go for walks, as evidenced by body language such as their tail being up, but others spoke of dogs that only liked to walk in good weather or that were fearful on walks and hence were not taken for walks often.

"One of the nice things about dog walking is how we can benefit from it," said Westgarth. "There's something about dogs, and walking dogs, and my participants say it doesn't feel like exercise." She added, "You know you'll enjoy it and you know they'll enjoy it, so just put the lead on and go." And the size of the dog does not matter, she said: "Just because a dog's small, doesn't mean it can't do a long walk. Because that's another perception we build, that it's alright they're little, they can cope with not having big long walks. But if you actually do take them out for two hours, they love it."

Research with Canadian dog owners found a number of reasons for walking the dog. Giving the dog exercise, ongoing socialization, and bathroom breaks were all motivations, but people also appreciated the way it gave them a reason to be physically active.[7] Sometimes all of the family walked together because the dog liked it. Sometimes friends and neighbors were also involved in dog walks or would care for the dog (e.g., if the owner was on vacation). People found ways to give their dog opportunities to be off-leash and changed walking habits in relation to the dog's needs. People said they felt a responsibility to look after their dog and to ensure they did not trouble other people.

There is some evidence that people generally take the dog's exercise needs into consideration. In an Austrian study, 61 percent of large dogs had a daily walk of at least forty-five minutes in a green space such as a park, compared with 50 percent of small dogs.[8] When researchers compared the amount of exercise the Kennel Club says each breed needs with the amount owners said they walked their dog, they found a positive correlation, with breeds needing more exercise generally getting more exercise.[9]

At least half of dog walks start just outside the front door and involve walking in the local streets. So it's not surprising that aspects of the neighborhood may also influence how much people walk their dogs.[10] Good sidewalks, good lighting, a feeling of safety, nice green spaces, off-leash dog parks, shade for the summer months, and washrooms (for the humans) are all on the plus side. When it comes to parks, those dog owners who have a park within 1.6 km (1 mile) of their house and that is a good place to walk the dog are more likely to spend at least ninety minutes a week walking their dog. From your dog's perspective, if there is no such park in your

Little dogs need walks too.
BAD MONKEY PHOTOGRAPHY

immediate neighborhood, it could be worth finding one that is a short drive away.

"ONE THING I think would make the world better for dogs is if they all got taken for an off-leash walk every day. There is no greater joy as a dog owner than to see butt tucked under, head back, doing zoomies across the grass. So many dogs don't get this opportunity. Perhaps their owners don't have access to a suitable and safe environment to walk them in this way, or don't have enough time in their busy schedules. Perhaps they are struggling with their dog's behavior on walks and having to confine them to a lead, a vicious circle. Most worryingly to me, many dogs are incapable of running around due to the conformation they have been bred with, or most commonly the fact that they are overweight. In order to run around like this and experience this joy of what it is to be a dog, fitness and lean body condition are key. We often think about obesity being caused *by* lack of exercise, but it is also a massive cause *of* lack of exercise. When I worked as an Assistance Dog trainer, we sometimes had to manage dogs through a weight loss program. It was amazing seeing both the physical and mental transformation from pudgy plodder to lithe racer. If my wish could be achieved, both dogs and their owners would live much happier lives."

—CARRI WESTGARTH, PhD, research fellow, Institute of Infection and Global Health, University of Liverpool

WHY LETTING THE DOG OUT INTO
THE YARD IS NOT A WALK

ONE REASON SOME dogs don't get walked is because their owners think the dog will get enough exercise while hanging out in
the yard. But it turns out dogs are inactive for most of this time.
Researchers in Australia took video of fifty-five young Labrador
Retrievers in suburban backyards over a forty-eight-hour period.[11]
Of those dogs, 52 percent got a daily walk of between thirty and
sixty minutes, but 31 percent were not taken for a walk at all. On
average, 74 percent of the time spent in the yard was inactive,
ranging between 45 and 96 percent.

The dogs in the study were more active and played more when
a person was in the yard with them, though they did sometimes
play with objects on their own. A dog was more active when more
than 1 percent of the yard was foliage (as opposed to a flat expanse
of grass); when the dog was kept inside the house at night and also
had a dog house in the yard; when the dog was reported to obey
commands when given; and when the dog moved around from
one door to another, to a window, to the gate, and back again
trying to keep tabs on their owner in the house.

In this study, problem behaviors (barking, chewing, digging,
and carrying or manipulating objects) were more common in the
yellow Labs than the brown and black Labs, in dogs who had not
been trained, and in dogs who were more active. Since some of
the activity was related to transitions from one place to another,
apparently to check on their people, it is possible some of the
problem behaviors were related to the dog not liking being apart
from their people.

This study shows that the design of a yard will affect what dogs
do there. Putting the dog out in the yard is not a substitute for a

proper walk, but having foliage gives dogs something to explore and may also bring birds and wildlife for the dog to observe.

DOG-RELATED ISSUES THAT AFFECT DOG WALKING

SOMETIMES PEOPLE DON'T walk their dog at all, or as often as they should, because of dog-related issues. Having a reactive dog who barks, lunges, or growls at other dogs or passersby; being concerned about other dogs in the neighborhood; and dealing with a dog that pulls on-leash so much that walking feels difficult are just a few of these reasons.

No words are more likely to strike fear into the heart of the owner of a reactive dog than "He's friendly!" It means someone is about to let their dog run right up to yours, even though you are trying hard to stay at a safe distance. This can undo weeks of hard work in an instant. We all make mistakes or find ourselves in a situation where our dog's recall isn't as good as we thought, but unfortunately some of the "He's friendly!" crowd are oblivious to the idea that others don't want their hairy, muddy dog to jump and slobber all over them and their dog.

When a dog is reactive on-leash, it can be really hard to tell if the dog is friendly to other dogs and just frustrated because they can't get near to them and play, or if they are actually terrified and desperate to keep the other dog away. For the friendly dogs, lots of off-leash play opportunities can help, because dogs are sociable creatures and enjoy hanging out or playing together. But it's important to keep fearful dogs safe, which means keeping distance between them and the other dog. And at the same time, you can teach them to like other dogs after all by making something nice happen, like feeding them pieces of steak, when another dog is in sight.

Walks with a reactive dog require stealth and planning: it means having your eyes and ears open for other dogs at all times, and being alert and ready to move behind a car or tree or change direction to keep the other dog out of sight. For some reason, this avoidance behavior sometimes offends other dog owners, and then you have the added issue of keeping your fearful dog away from a shouting person trying to advance with their "friendly" dog while giving completely wrong advice, like "Just let them work it out." Try to stay calm, and do your best to move away from the kind of experience that made your dog fearful in the first place.

Leash laws often govern where dogs should be on-leash and where it is optional. To study the effects of the leash, Dr. Westgarth recruited ten dog owners to walk a particular path with the dog on-leash and off-leash.[12] Again, interactions with other dogs were more common than with people. Using a leash reduced the number of interactions, but the big difference was if the other dog was also on a leash. So if you want to stop dogs from interacting with each other, they both need to be on-leash. Although dogs have more freedom to do doggy things off-leash, there are times when a leash is essential.

Unfortunately, sometimes dogs do attack other dogs. The consequences can be severe not only for the attacked dog but also for that dog's owner, who will find it stressful even if they are not physically hurt themselves. A particularly severe example is when guide dogs (seeing-eye dogs) that provide services to blind or partially sighted people are attacked. The Guide Dogs for the Blind Association in the UK detailed 629 attacks over a fifty-six-month period from 2010 to 2015.[13] The results showed an impact on working ability of the dogs (some of whom had to take time off work or be withdrawn from the program), injuries to the handlers and other people, and effects on the handler's emotional well-being and mobility. Perhaps the most shocking part of this study is that the owner of the attacking dog was present 76.8 percent of the time. The researchers point out that if they had put their dog on-leash as soon as they saw the guide dog's harness, many of these attacks could have been prevented.

Another issue people sometimes have is that walking their dog is difficult because the dog pulls. Some people use a choke or prong collar to stop their dog from pulling, but this is an aversive technique (see chapter 3). A choke collar tightens unpleasantly when the dog pulls, releasing only when they stop pulling. A prong collar has sharp prongs that dig into the dog's neck when they pull. Sometimes people try one of these collars on their arm or neck before deciding it is okay to use on their dog, and I think you have to give them credit for trying it. The trouble is that, unlike the dog, people typically have control over the collar's use, which makes a big difference. Also, the San Francisco SPCA notes that "[the] skin on a human's neck is actually thicker (10 to 15 cells) than the skin on a dog's neck (3 to 5 cells)."[14]

We tend to think that because dogs have fur, their skin must be more protected than ours. But a dog's neck is a very sensitive area: it contains essential body parts like the windpipe. Applying pressure to the windpipe is not good for any dog, but can be especially serious for brachycephalic dogs that already struggle to breathe. The other thing to remember is the way choke and prong collars work. If they stop a behavior, it must be because the animal finds the collar aversive.

It's better to train your dog to walk nicely on-leash using rewards, and to use a no-pull harness as a management tool if needed. A no-pull harness has a clip at the front (as opposed to a harness with the clip at the back, which is suitable for dogs that don't pull). A team of scientists investigated the effects of walking a dog on a harness and on a regular neck collar by walking the same dogs using one and then the other on separate occasions.[15] Their behavior was monitored for signs of stress. The results showed that harnesses do not cause stress and are a great choice for walking your dog.

Dr. Tamara Montrose, one of the authors of the study, told me in an email that "Whilst neck collars are widely used when walking dogs, concerns have been raised about their potential to damage the neck and trachea. Furthermore, collars can be problematic in dogs with eye conditions such as glaucoma. Harnesses are often anecdotally proposed to be better for dog welfare." In their study, she continued, "We found that there were no differences in behavior between dogs walked on either a neck collar or a harness... Our findings suggest that dog welfare is not compromised by either form of restraint; however, we are interested in undertaking future study with a range of different brands of harness and collar, consideration of physiological stress indicators and assessment of gait and magnitude of pulling."

The bottom line is that while many people are doing a good job of giving their dog enough exercise, others are not. Walks are important to give dogs exercise, sniffing opportunities, social interactions with other people and dogs, and the ongoing socialization those experiences bring. Recognizing these benefits to dogs can help motivate you to take your dog for walks. One thing's for sure: Once you get into a routine, your dog will remind you when it's time for your walk!

HOW TO APPLY THE SCIENCE AT HOME

- Go for walks. They are good for your dog—and good for you too. If you are not in the habit, try to develop a routine—your dog will soon learn it and come to remind you when it is time! Whether it's one or two walks a day (not counting toilet breaks) depends on you and your dog.

- Use appropriate gear (e.g., a no-pull harness) and train your dog to walk nicely on-leash and to come when called.

- If you need help to resolve problems such as fear or reactivity, hire a qualified dog trainer. Think about ways to manage the problem while you work on it (e.g., avoiding places where dogs are off-leash or walking at quiet times of the day).

- If you see someone trying to keep their leashed dog away from you and your dog, cut them some slack and give them space; they may be trying hard to keep a reactive dog feeling safe. Insisting that your dog meet theirs or offering unwanted training advice will not be welcome or helpful.

- If dogs must be on-leash in your neighborhood, get creative to find off-leash opportunities such as dog parks, fenced fields, tennis courts for hire, or a friend's fenced yard.

10

ENRICHMENT

·················

O NE DAY LAST summer I awoke to a strong smell of wood smoke. I got up and looked out of the window; smoke hung visibly in the air. A fire truck drove slowly up the street as if looking for something, and shortly after, it came round again. It turned out several other people in the neighborhood had smelled the smoke too and called 911 thinking there was a forest fire nearby. But the wildfire was far away, the smoke blown in on the wind. It hung around for days, making long walks danger-ous for both dogs and humans. Our trips outside were limited to canine bathroom breaks. How did we entertain Bodger indoors? We fell back on an old favorite: tug.

Tug is a daily ritual for Bodger. When he gets excited, he will run and get his rope and ask hopefully for a game; when we're watching TV, he will unceremoniously dump the rope in my lap, and then sit and stare at it. When he wins a game, he does a vic-tory lap of the living room with the rope dangling from his mouth.

People used to argue that playing tug with dogs caused behav-ior problems. So researchers played tug with fourteen Golden Retrievers to find out whether or not it changed their behavior.[1]

Each dog played tug with the experimenter forty times. Half of the time they were allowed to win, and the other half they lost, in case winning or losing made a difference. Before and after, the dogs were tested on their obedience and reactions to the experimenter. The range of tests included calling the dog, asking them to sit, and some things you shouldn't try with your dog at home: forcing the dog into a down position and taking their food bowl away to see what they would do.

The findings were good news. Dogs were more obedient and attentive after play sessions, whether or not they were allowed to win. Winning didn't lead to any behavior issues. In fact, the dogs were more involved in the game when allowed to win, so it's a good idea to let your dog win from time to time!

THE IMPORTANCE OF ENRICHMENT

ENRICHMENT MEANS MAKING changes to the environment that will improve your dog's welfare. It is good for their mental health, because enrichment encourages them to interact with their environment and engage in normal canine behaviors. If we don't give dogs these opportunities, they will often find them for themselves, like when they chew household items. Chewing is a normal canine behavior, but sometimes dogs don't have a way to express it that is permissible to their owner. The solution includes making sure the dog does have something they can chew, and luckily there are plenty of chew toys to fill the void.

We often think of enrichment when it comes to zoo animals: watermelon being given to captive wolves to eat on a hot day, or trees and other climbing spaces being added to zoo enclosures. We can do this with our dogs too. Activities that stimulate our dogs' senses and challenge them to solve problems are important for their happiness. One measure of successful enrichment is whether the animal chooses to interact with the item or not. Other ways of evaluating success would be quality-of-life measures or observations of the animal's behavior.

A study reported in *Animal Cognition* investigated whether dogs enjoy the experience of solving a problem in order to obtain a reward, or if it is just the reward itself that makes them happy.[2] Rather unusually, the idea came from a study that found cattle that completed a task to earn a reward seemed to be happier than those that just received the reward. The study involved six matched pairs of Beagles (twelve dogs in total). Each animal was an experimental dog for half of the time and a control dog for the other half. There were also six pieces of matched equipment, including a dog piano that had to be pressed to play a note, a plastic box that had to be pushed off a stack so it would noisily hit the floor, and a paddle lever that would make a bell ring. When manipulated correctly by the dog, each piece of equipment made a distinct sound that showed the task was complete. That opened a gate to a ramp that led to a reward.

Each dog was trained on what to do with three of the pieces of equipment and their matched pair was trained on what to do with the other three. In the experiment itself, one piece of equipment was present. When the experimental dog performed the behavior they had been trained to do, the gate opened to give access to the ramp leading to the reward. Subsequently, when it was the control dog's turn, it did not matter what they did; the gate opened after the length of time it had taken the paired dog to solve the puzzle, and both dogs got the identical reward. In other words, the only difference between the experimental and control conditions was whether or not the dog's manipulation of the equipment had an effect on the gate opening.

The scientists observed that the experimental dogs were keen to get to the start arena and usually went into the room ahead of the assistant. On the other hand, the control dogs became reluctant and often had to be coaxed to enter the room.

Dogs in the control condition were less active in the start arena. They would sometimes bite or chew on the equipment, which dogs never did in the experimental condition. Once the gate had opened, the control dogs were quicker than dogs in the experimental condition to enter the runway and leave the start arena. There were no differences in mean heart rate, however. Dogs in the experimental condition wagged their tails more, which also suggests they were happier. Dogs in both conditions were more active when expecting a food reward, which is consistent with other studies that found dogs prefer food to petting.

Towards the end of the study, some of the dogs were successfully manipulating the item they had *not* been trained on, though of course it had no effect on the gate. This study shows that having control over a situation and being able to solve problems is good for dogs' welfare.

In an email, Dr. Ragen McGowan, first author of the study, elaborated on these findings: "It has long been our impression that our pets have rich emotional lives and that their experiences affect them profoundly in ways similar to how humans are affected. We are now starting to be able to back this up scientifically, which is very exciting. Think back to last time you learned a complicated new task... do you remember the excitement you felt when you completed the task correctly? Our work suggests that dogs may also experience this 'Eureka Effect.' In other words, learning itself is rewarding for dogs." She added, "Providing your dog with opportunities to solve problems (e.g., cognitive puzzle toys, or a game of hide and seek with treats in your yard) or learn new behaviors can be quite rewarding for your dog. Many pet own-ers understand the importance of keeping their dogs physically active; our research helps to emphasize the importance of keeping dogs mentally active as well."

Research shows that small dogs tend to miss out on both training and play opportunities with their owners.[3] Of the vari-ous activities asked about, only agility (a dog sport in which the handler guides the dog around an obstacle course including weave poles, tunnels, and teeter-totters) was done equally by both small and large dogs. Smaller dogs missed out on training, games of tug, tracking and nose work, and going jogging or cycling with their owner. This is one of several reasons why small dogs are less obe-dient than large dogs. So remember that whatever size your dog, they need some kind of enrichment.

THE VALUE OF SENSORY STIMULATION

ONE WAY TO think about enrichment is in terms of a dog's senses. It's important to note, though, that what is important to a dog is not the same as what is important to a human. A dog's vision is 20/75 compared with our 20/20, meaning that what we see at 75 feet (23 meters), the dog can only see at 20 feet (6 meters).[4] They see the world in yellows and blues, rather like a person with red-green color blindness (the bright colors of dog toys are more for the benefit of humans than dogs).

A dog's hearing range is broader than humans'. Whereas humans are said to hear a range from about 20 to 20,000 hertz, dogs hear much higher frequencies, from about 67 to 45,000 hertz.[5] That's the reason behind the design of those high-pitched dog whistles, some of which can be heard by dogs but not by humans. Incidentally, this means some high-pitched whirrs or hums from electronic equipment in the house may be audible and possibly annoying to your dog; similarly, the flicker rate of some lighting or the artificial scents of some household products may affect dogs differently from humans too.

There are some suggestions that music (at least certain kinds) may be relaxing to dogs. A study of kenneled dogs found they slept more when classical music was played compared with heavy metal, music designed for dogs, or no music at all.[6] The dogs shook more during heavy metal and vocalized less during classical music compared with no music. Since only a few tracks of each type of music were used, the results may not generalize to genres as a whole. Subsequent research by a different team found that shelter dogs were more relaxed when listening to classical music, but they also got used to it.[7]

"LARGE NUMBERS OF dogs are kept in rescue shelters, which are often stressful environments for the residents. I believe that greater use of sensory enrichment, such as classical music and audiobooks, would be beneficial for shelter dogs to reduce stress and potentially improve welfare. Such enrichment also tends to be easily applied and relatively inexpensive which is an important consideration in shelters. In addition, using sensory enrichment that is not only beneficial to dogs but also appreciated by visitors may have the potential to encourage them to spend more time at the shelter and potentially help adoption rates."

—TAMARA MONTROSE, PhD, principal lecturer, Hartpury College

A dog's most important sense is smell.[8] Dogs don't just have a nose. They also have something called a vomeronasal organ (VNO), which is tucked away in the upper palate, and the ability to suck molecules in and out and in again to increase their availability for detection. The VNO is not open to airflow from the nose; instead molecules must dissolve, such as in the saliva, from which they can enter the VNO through two ducts just behind the front teeth. Have you ever been disgusted to see your dog lick urine or feces? They may have been making information available to the VNO. Similarly, we see dogs checking messages left in urine all the time; sometimes they even sniff us more than we would like, as with an embarrassing nose against the crotch or bottom.

Dogs have an amazing sense of smell, and Dr. Alexandra Horowitz, canine scientist at Barnard College in New York, says they can even smell the time of day due to the way changes in temperature affect scent molecules. Because of their acute sense of smell, dogs are used professionally as medical detection dogs, drug detection dogs, even *C. difficile* detection dogs. In the related recreational sport of canine nose work, dogs are trained to search for a scent. At the beginning, the arena is a set of boxes spread out across the floor. The dog is allowed to sniff around the boxes until they find and eat the hidden piece of food. Over time, the dog is trained to search for a particular scent, such as sweet birch or anise. As dogs progress through the sport, the locations become more complex, such as cars or buildings, maybe even a sweep through a disused school to find the target smell, just like a professional detection dog would do. One study found that dogs who took part in a nose-work class were more optimistic than those who instead participated in dog training (learning to heel for rewards).[9] The conclusion is that activities involving autonomy and the sense of smell are good for dogs' welfare.

"Dogs love nose work because we are playing with them in their world," said Ann Gunderson, a dog trainer who has earned many nose-work titles with her Australian Cattle Dog and Nova Scotia Duck Tolling Retriever, including the first Triple Odor Games (TOG) dog and the first Masters TOG dog at the K9 ABC Games championship. "What the titles represent is me being a team member with my dog," she said, "and with nose work that is the best feeling ever. I'm watching my dog and listening to what she's telling me." She explained, "Dogs really care about what we think. You know, our pet dogs, they watch us 24/7 and in most cases they're trying to figure out how to make us happy. Or perhaps just to get treats, because they're dogs!" Humans are always trying to get the dogs to do something, like sit or stay, she noted. "These are all things we ask the dog to do for us. And with nose work the dog is doing it for themselves, with us tagging along. They get to be the experts, and what I see is that it's pure magic for them, they love it so much."

"LET THEM SNIFF. Perhaps because we humans are so visually centered, it's hard for us to imagine what it might be like for our primary sensory ability to be olfaction. But that's how it is for dogs: they sniff first, and ask their eyes to confirm or deny. Their world is made of scents more than sights. As a result, when they agreeably head out with you for a walk, the two of you are experiencing parallel universes: we see what's on the street; the dog smells who's passed by and who is upcoming (on the breeze). Since humans are generally averse to closely smelling things—in fact, we find the idea of 'smelling' one another funny or even rude—some owners discourage dogs from doing that—from sniffing one another or the traces other dogs have left. But that is the dog's whole world. I would no more pull my dog away from a street corner he is mightily investigating than I would force my son to stare at his knees as we drive by the Colosseum. Acknowledging the dogs' otherness—and in this case, their different way of perceiving the world you share—is a good step toward giving them the life they deserve."

—**ALEXANDRA HOROWITZ**, PhD, adjunct associate professor, Barnard College, and author of *Inside of a Dog: What Dogs See, Smell, and Know*

One of the nice things about K9 Nose Work is that the approach to teaching dogs is wholly positive and kind. It's also suitable for reactive dogs, because only one dog is in the arena at a time. Gunderson told me she has known brachycephalic dogs, blind dogs, deaf dogs, tripod dogs, and elderly dogs who take part in and enjoy nose work. "Any dog can do nose work," she said, because they are allowed to do it in their own way. She also said the benefits extend beyond the nose work session itself, because of the way handlers must learn to observe their dog so closely. The result is that handlers become more aware of what their dog likes, and in particular how much they like to sniff. "People will start to walk their dogs differently; they'll actually let them sniff, and that's lovely," she said.

DOG SPORTS AND THE HUMAN-ANIMAL BOND

THERE ARE ALL kinds of ways to give dogs the chance to problem-solve, including reward-based obedience training, doing agility, learning tricks, playing with puzzle toys, doing nose work, and many other enrichment activities. People can participate in dog sports at any level, from local beginner classes to national and international events. The types of sports are varied, so people and their dogs can find the right match for them.

Although there is a chance to win titles, it turns out this isn't the main reason why people take part in dog sports. According to a study published in *Anthrozoös*, people who participate in a lot of dog sport events score highly on levels of intrinsic motivation, relating to internal desires and interests, as well as extrinsic motivation, relating to external factors such as rewards and titles.[10] The statistical results were supported by comments from participants. "When you work with the dog as a team in a sport, you and the dog develop a very special relationship," said one. Another said, "I like the connection that develops with a dog during training and I like being around people who feel as I do about dogs. It enables me to connect with people from all walks of life." Other participants also clearly put the dog at the center of the experience. "I enjoy the time I spend with my dogs and the friends that I have made over the years because of the dogs," said one. Many participants mentioned the connection with their dog and physical exercise for the dog as important motivators. Others cited the opportunities it gives them for learning. One person said it was "fun to train—fun to think outside the box and find new ways to teach with different dogs over the years—love 'light bulb' moments." It's nice to think that both the dogs and their owners are getting many "light bulb" or "eureka" moments through practicing for and competing in the sport.

For dogs, like any athlete, there can be a risk of injury arising from dog sports. In some cases dogs may also find competitions stressful; in a small study of seventeen agility dogs, scientists found some signs of stress, such as restlessness, before and after the event.[11] As with anything, it's a case of needing to take part in moderation, and keeping an eye on your dog to ensure the experience is positive for them.

HOW AND WHY TO ADD ENRICHMENT

THE EFFECTS OF enrichment have been studied more in dogs in kennels—either dogs who live permanently in kennels or those in rehoming centers—because those environments are already impoverished, meaning they are less stimulating for dogs than the average home. However, once you take into account the fact that many people go out and regularly leave their dogs home alone, it becomes obvious that ongoing enrichment is a good idea for any dog.

Dogs like new things, a preference known in scientific terms as neophilia. The corollary is that they can lose interest in (habituate to) familiar toys over time. One study tested this effect by repeatedly presenting shelter dogs with the same toy.[12] By the time they had been presented with a toy ten times, most of the dogs were no longer interested in it. But the researcher also found that changing the color or the scent of the toy (or, best of all, both) led dogs to become interested again. Even a small change is enough to make things seem new, which means you can make the old toys more interesting again by playing with them yourself or by bringing them out on rotation.

Another study, this time of dogs that live in kennels at a lab, also found that giving them a Kong dog toy stuffed with food led to dogs (not surprisingly) eating from the toy and not habituating to it.[13] As well as interacting with the toy, they were more active in general and barked less. A study of dogs at a rehoming center found they had an improved quality of life if they were provided with enrichment.[14] Exercising for more than thirty minutes a day, training for thirty minutes a day, being given a combination of wet and dry food (instead of just dry), and having a quiet environment all led to much better quality-of-life scores.

The bottom line is that lots of enrichment is ideal, but you can start with simple ideas and go from there. For example, I like to play a game with Bodger where I show him I have a treat and then put my hands behind my back and swap the treat from one to the other. Then I bring my hands out in a fist for him to pick one. If he picks the empty hand, he gets nothing. But if he chooses the hand with the treat in it, he gets the treat. Then we play again.

Another free option is simply to allow your dog as much sniffing time as they like on walks—turning the walk into a "sniffari." Instead of hurrying along, wait patiently while your dog catches up on the "peemail." Sniffing is good for them because it's an important way for them to get information about the world, such as which dogs have been by and what they've been eating. And maybe while your dog is sniffing the things we find disgusting, you can turn your attention to the bees on the bee balm, the creamy color of the clover in the grass, or little clouds scudding across the sky.

Remember that you don't have to be perfect. The best approach is to pick one thing to provide enrichment. Try it and see how your dog likes it. Evaluate it like a scientist and ask, is the dog using the enrichment item? Are there any changes in their behavior as a result? Make a few tweaks if necessary to ensure the level is right for your dog. If they like it, keep on doing it. Then after a while, pick another thing and add that in. This is the easy way to make enrichment a habit. Your dog will be happier for it.

HOW TO APPLY THE SCIENCE AT HOME

·   Try to find ways to let your dog use all of their senses.
Since the sense of smell is so important to dogs, give them
opportunities to use their nose, such as taking "sniffaris" in
which you let the dog's nose dictate where to walk. Hide food
in snuffle mats (available from pet stores, or you can make
your own) or scatter food in the grass in your yard. Make a
beginners-style nose game in your family room by spreading
some cardboard boxes around, hiding a piece of food in one of
them, and then letting your dog into the room to hunt for it.
Reward their success by adding another piece of food to the
box.

·   Use reward-based training to exercise your dog's brain and give
them many chances to earn food. If your dog already knows
basic obedience, try teaching tricks or Rally-O (Rally Obedience, available in classes and competitions). There are lots of
great YouTube videos that show you how to teach tricks like
"play dead."

·   Ensure your dog has chew toys (because dogs like to chew),
food toys (to make them work for their food), and other toys
so they can play games with you (like fetch or tug). Toys provide enrichment and satisfy your dog's needs to engage in
normal canine behaviors like playing and chewing. Make sure
the toys are safe, and if necessary put them away when you're
not there (e.g., so that squeakers won't be eaten).

·   Choose dog sports that you and your dog will enjoy so it's
fun for both of you. For example, if your dog loves to run and
you want to get fit, you could try cani-cross (cross-country
running with your dog attached to you by a harness so you

run together as a team). Look for opportunities to try out dog sports, or observe a competition or class.

- Remember that while dogs like brand-new toys, having familiar toys on rotation and/or washing them can be enough to make them new again.

- Consider your dog's living environment and provide things like dog beds to give them a relaxing place to sleep, a sandbox in the garden to allow space for digging, and so on.

11

FOOD AND TREATS

....

MEALTIME IS BODGER'S favorite time of day. While his food is being prepared, little drops of saliva fall from his mouth onto the lino floor. Soon after he starts to eat, he always looks back at me and I feel like he is saying he is grateful for his food. If it's not his mealtime but mine, he is hopeful of getting a little morsel of something. If my husband is eating snacks, he is very hopeful indeed. And if it's the cats' mealtime... well, I had to train him to stay at a distance because he used to watch so closely in case of leftovers. The cats are carnivores and must eat a meat-based diet. No wonder Bodger likes their food! I rarely eat meat, but he likes my food too. And a piece of raw carrot or the end of a zucchini is a prize to be taken away and nibbled on in peace.

Good nutrition is one of the welfare needs of dogs, and it turns out that the food we give our dogs relates not just to nutrition and beliefs about what dogs need but also to an expression of the dog being part of the family.

WHAT DOGS CAN EAT: THE EVOLUTION OF MODERN METABOLISM

DOGS ARE OMNIVORES and can eat a range of different types of food. The ability to digest starch, such as rice and potatoes, is one of the ways dogs differ from modern-day wolves. Although it makes perfect sense in hindsight that dogs have adapted to co-exist with us in this way, it was an unexpected finding in a study by Dr. Erik Axelsson in Sweden.[1]

To understand the genetic changes that accompanied the transformation of ancient wolves into domestic dogs, Axelsson and his colleagues sequenced the DNA of twelve wolves and sixty dogs of fourteen different breeds. They focused their research on regions of the DNA where there was little variation in dogs, which suggests those parts were so important for the survival of domesticated dogs that any variation there was lost. Some of these regions contain genes related to brain function. The unexpected part was that some of these regions contain genes related to the digestion of starch. Dogs have between four and thirty copies of a gene, Amy2B, for a protein called amylase that starts the break-down of starch in the intestine. Wolves only have two copies of this gene. Having more copies of the gene means dogs have more amylase than wolves, and tests in the lab showed dogs should be five times better than wolves at digesting starch.

Another gene called MGAM, related to a protein called maltase, is also important in digesting starch. Dogs and wolves have the same number of copies of that gene, but there are key differences. Dogs make a longer version of maltase, which means it is more efficient. The longer version is also found in omnivores and herbi-vores and seems to be important for digesting plant protein. With these findings, Axelsson proposes that as wolves—not modern-day wolves, but their ancestors—began to hang out around human settlements, they developed the ability to digest starch. They would have been able to get food that wasn't meat left over by humans, so being able to digest it became an important survival advantage.

A further survey of different types of dogs from around the world found most wolves, coyotes, and jackals also have just two copies of Amy2B.[2] This was also the case for two types of dog from the Arctic and Australia—Siberian Huskies and dingoes—who, until recently, lived with people who mostly ate meat. The other dog breeds studied had more copies of the gene, just as in the earlier research. Another study found variation in the number of copies of the Amy2B gene amongst different dog breeds, with Greenland Sledge Dogs (based in Greenland) and Samoyeds hav-ing amongst the lowest number.[3] The scientists could not rule out recent interbreeding between wolves and the Greenland Sledge Dogs, however. They also found some variability within breeds and noted that although Amy2B has a big influence on the diges-tion of starch, it is not the only factor.

We still don't know when domesticated dogs developed the ability to digest starch, but researchers looked at the genome of ancient dogs and wolves in order to find out.[4] They extracted DNA from the teeth and bones of thirteen dogs found at archaeolog-ical sites in Europe and Asia. These ancient dogs had from two to twenty copies of Amy2B, showing that during early farming not all dogs had gained the extra genes. But it also shows that selection for this gene started at least 7,000 years ago. These fascinating studies suggest dogs and humans have both evolved genes to help them digest starch as a result of farming. And it also shows the commonly held belief that dogs should be fed like wolves is not correct. Somewhere in the past, the common ancestors of today's wolves and dogs diverged to enable dogs to eat a much more human-like diet.

DOGS AND DIET

OUR OWN CONCERNS about diet and health are reflected in what we feed our dogs. A survey of dog owners in Australia found most people (41%) feed their dog twice a day, about one-third (36%) feed once a day, and a few (16%) leave food down all the time.[5] Most dogs were given treats once a day or once a week (37% for each). The most common foods given were medium-priced kibble (44%), raw bones (44%), and kitchen scraps (38%). A similar survey of owners of Labrador Retrievers in the UK also found kibble was the most commonly fed food, given by 80 percent of owners, while 13 percent gave their Lab a mix of dry and wet food.[6] While puppies were fed three or more times a day, by the time the dogs reached 6 to 9 months, most were fed twice a day.

Commercially available kibble and canned pet foods are manufactured to meet specific nutritional requirements (including some for different life stages or for special diets), and are available for every price point. Manufacturers spend time ensuring the texture, smell, and shape of kibble will be attractive to dogs. Some people believe feeding kibble is good for a dog's teeth, but there is a lack of evidence to support this idea (with the exception of veterinary dental diets).

Some dog owners prefer to feed their dog a different kind of diet. A paper in *Veterinary Clinics of North America: Small Animal Practice* looked at the reasons and whether there are advantages or disadvantages to these diets.[7] Some people choose to feed their dogs a home-cooked diet because they want to avoid additives (such as preservatives) or animal by-products not used in human foods or because they want to use organic food. Others feel that preparing food for the dog helps them feel the dog is part of the family. But analyses of recipes for home-cooked dog foods show many of them are lacking in certain ingredients.[8]

Some people may cook in order to give their dog a vegetarian or vegan diet, but again it is difficult to get the nutritional content right. Studies of commercially available vegetarian dog foods found many issues with nutrition and suggest it is important to pay attention to quality.[9]

Feeding a home-prepared raw diet is also hard to get right from a nutritional perspective, just like a home-cooked diet, but an increasing number of commercial raw diets are available. Some people feed a raw meat-based diet because they think it is closer to what dogs would eat "in the wild" or in their evolutionary past, while others believe it is better for their dog's health or teeth. Unfortunately, there is a lack of clinical trials on the effects of feeding a raw diet, so there is no documented evidence of its benefits. Raw meat-based diets typically have a higher amount of dietary fat, which may lead to a glossier coat but may also cause gastrointestinal upset and weight gain for some dogs.

There is only limited research on the nutritional content of raw diets. Of more concern is the risk to the dog's health from bacteria in the food, and a subsequent risk to people in the house if dogs shed those bacteria in their feces, which can happen without them showing signs of being ill.[10] One study tested commercially available raw meat-based diets for four types of bacteria and parasites, and found the bacteria *Listeria* in 43 percent, *E. coli* in 23 percent, and *Salmonella* in 20 percent.[11] There was also a risk from parasites if raw meat had not been frozen. The scientists involved in this study say dogs (and cats) fed a raw meat-based diet are more likely to get infected by antibiotic-resistant bacteria, which is a risk to their health as well as to humans. They suggest that labels on raw meat-based foods should give guidance on storage and handling.

If feeding a raw diet, follow the same food hygiene practices as you would with human food (e.g., raw chicken).[12] Keep cooked and raw foods separate, defrost foods in the fridge on a low shelf, keep pet food bowls separate, and wash your hands after feeding your dog and picking up poop. Remember that some snacks and treats given to dogs are also raw, such as pigs' ears, bully sticks, and some freeze-dried treats. The risks are greater for vulnerable members of your household, such as children, seniors, and anyone with a compromised immune system.

Human foods that are not safe for dogs are shown in the following list. Some foods, such as chocolate, caffeine, and xylitol—which is found, for example, in low-calorie peanut butters as a sweetener—are toxic to dogs. Large amounts of fat can cause pancreatitis. As mentioned above, raw or undercooked meat, eggs, and bones can be a risk for bacteria. Some foods (such as yeast dough) can cause bloating, while other foods (such as cooked bones, pits from fruits) can cause an obstruction. Some foods, like coconut, are safe in small amounts. Many dogs lack the enzyme needed to digest dairy and so are lactose-intolerant. While most nuts are not safe for dogs, peanuts are okay (but remember to check peanut butter for xylitol!).

Human foods that are not safe for dogs

- Alcohol
- Avocado
- Chocolate (dark chocolate is most toxic)
- Citrus fruits
- Coconut (small amounts unlikely to cause upset)
- Coffee (and other sources of caffeine)
- Grapes and raisins
- Milk and other dairy
- Nuts (including macadamia nuts, almonds, pecans, walnuts)
- Onions, garlic, and chives
- Raw or undercooked meat, eggs, and bones
- Too much salt
- Xylitol (artificial sweetener used in some peanut butter, candy, etc.)
- Yeast dough

Source: ASPCA Poison Control[13]

THE TRUTH ABOUT TREATS

GENERAL GUIDELINES SUGGEST treats should not be more than 10 percent of the dog's daily calorie intake, but a study of dog treats published in the *Veterinary Record* found that, on average, for both small-and medium-sized dogs, the treats exceeded the recommended amount of daily energy intake, with the exception of dental sticks.[14] Sugars were a common ingredient, as dogs (unlike cats) can taste sugar. The lack of detail on the labels of treat packets was noted as a concern, and the scientists said treats should not be given to any dog who is on an elimination diet and who is therefore at risk of an adverse effect such as an allergic reaction. As well, the analysis showed many treats were not suitable for dogs with chronic heart failure or chronic kidney disease. Check with your vet about which treats are safe to use if your dog is on a special diet.

THE PROBLEM OF OVERWEIGHT AND OBESITY IN PET DOGS

BODGER IS ALERT to all kinds of food-related noises: the fridge door opening, a plastic bag rustling, a potato chip crunching as it's eaten, the lid coming off the plastic tubs in which we keep cheese, a can being opened, the door opening for the cupboard where cat treats live, and the cat-treat packet rattling as it's shaken. Any food preparation brings him to the kitchen. He loves treats, he loves whatever table scraps we are willing to give him, and given the chance he is always willing to finish off what the cats don't eat from their meal. I have to keep a constant eye on Bodger's weight, and a few years ago we had to put him on a diet for a while.

Being overweight or obese is a common problem for dogs. Overweight is defined by a body condition score, which is measured on a nine-point scale (or on a corresponding five-point scale in which the points are reduced by half). You will find body condition score charts on the internet, and you can also ask your veterinarian. Dogs (and cats) are considered overweight if their body condition score is 6 or 7 out of 9 (corresponding to 10 to 20% more than their ideal weight), and they are obese if the score is 8 out of 9 (30% more than ideal weight).[15]

A large US survey put the rate of overweight and obese dogs at 34 percent, and an Australian study found 33.5 percent were

overweight and 7.6 percent were obese.[16] This doesn't necessarily mean dogs are fatter in one country than another, as the
data were collected at different points in time. But it does show
how widespread a problem it is. The US survey found overweight
dogs tended to be older, which makes sense as they had had more
time to gain weight. They were more likely to be fed semi-moist
food and to be spayed or neutered. Certain breeds were also more
at risk, including Golden Retrievers, Labrador Retrievers, and
Dachshunds.

Just as overweight and obesity are not good for human health,
they also put a dog's health at greater risk, particularly for musculoskeletal disorders and cardiovascular problems.[17] In the
Australian study, overweight and obesity increased with age up
to a point and then leveled off. While it's not clear why, one possibility is that the overweight and obese dogs became ill and did not
live as long, whereas their normal-weight counterparts carried
on with lower rates of disease. This study also found dogs in rural
or semi-rural areas were more likely to be overweight or obese.
There may be more sources of food available in these areas, or perhaps dog owners are less likely to take their dogs for walks and
more likely to assume the dogs will exercise themselves. Other
risk factors for overweight include being fed once a day (rather
than twice), being fed snacks, not getting enough exercise, and
having an owner who is overweight or obese too.

Obesity in young dogs is of particular concern, as overweight
puppies are likely to become overweight dogs. However, because
puppies of different breeds are going to grow into adult dogs of
different sizes and shapes, it can be hard to assess whether or not
a pup is growing normally or becoming overweight. Nowadays,
puppy growth charts for dogs up to 40 kg (88 pounds) are available from the Waltham Centre for Pet Nutrition to help your vet
monitor the weight of your puppy as they grow.

"THE DOG IS not set in stone. In the 21st century, we live off the internet. With the click of a button, we can find a plethora of information—as well as misinformation—about dogs. For the love of a dog, to quote Dr. Patricia McConnell, I hope that people continually take a step back. Try and notice everyday assumptions and expectations about dogs, however small, and then consider holding on to them a little less tightly. Our beliefs about dogs needn't be set in stone, because the dog is not a stagnant, all-known being. Create space to question assumptions and bring in new and ever-evolving information, particularly from researchers, veterinarians, practitioners, and science communicators who are doing the same. This, I believe, has the potential to enhance individual dog well-being."

—JULIE HECHT, PhD candidate, the City University of New York, and author of the *Dog Spies* blog at *Scientific American*

Why are so many dogs overweight? Our relationship with food plays a role since we are responsible for everything they eat. A German study published in the *Journal of Nutrition* compared sixty owners of normal-weight dogs with sixty owners of overweight dogs.[18] Both groups had an equally close relationship with their dog and both groups fed their dogs a commercially available diet. But the researchers found owners of overweight dogs talked to their dogs more, were more likely to let their dog sleep on the bed, were less likely to worry about contracting a disease from their dog, and did not see exercise or work as important for the dog. They also were more likely to say the cost of dog food mattered to them. Sixty-six percent bought food at the supermarket, compared with 47 percent of the owners of normal-weight dogs;

foods purchased from pet stores or from veterinarians are often of better nutritional quality.

What struck me most about this study was how much time some people say they spend watching their dog eat. Twenty-five percent of the owners of overweight dogs spent more than half an hour a day watching their dog eat compared with 11 percent of owners of normal-weight dogs. The overweight dogs were also fed more meals, snacks, and table scraps, and were more likely to have an overweight owner. This finding suggests that for some owners, using food to interact with the dog is particularly important. So, just feeding the dog less at mealtimes (and actually measuring the amount of food) alone would probably not make a difference. Changing the owner's behavior towards their dog—feeding them fewer table scraps and replacing some of the interactions that happen around food with petting, games of tug, walks—would also be necessary. Any food rewards used in training should be included in the dog's daily calorie ration.

The effect of people's attitudes on their dog's weight was also investigated in a study published in the *Journal of Applied Animal Welfare Science*.[19] Owners were asked to say whether their dog was underweight, a normal weight, or overweight, and then each of the dogs was assessed independently to determine their actual body condition score. Perhaps most worrying was the finding that many owners of overweight dogs did not know their dog was overweight. Of course, many owners who knew their dog was overweight had not succeeded in getting their dog back to a normal weight either. Lack of knowledge about an appropriate amount to feed the dog, as well as feeling that this measure was not important, was one important contributing factor. Another reason was owners feeling a lack of control over how much they fed their dog, which may mean they find it hard not to give in when the dog is begging. It has to be said that dogs are very good at begging for food!

One response to overweight and obesity in dogs is to put them on a special diet, many of which have been shown to benefit weight loss in controlled conditions. However, it also seems to be important to target owner behavior. A review of the literature published in *Preventive Veterinary Medicine* looked at interventions to help owners help their dogs lose weight.[20] The interventions were effective in changing dogs' body condition and owner behavior. Some of the techniques that were designed to change owner behavior:

- Setting goals for behaviors. For example, deciding to walk the dog for a certain length of time every day, or deciding on a fixed number of treats to give the dog each day so they are not overfed.

- Setting goals for outcomes, such as a certain amount of weight loss per week.

- Increasing the owner's knowledge about what and how much to feed the dog and how much exercise the dog should have.

- Setting strategies to monitor behavior. For example, using signage to show when the dog has been fed so they don't get fed extra times.

- Giving regular feedback by having the dog visit the vet regularly to be weighed.

Overweight and obesity put dogs at greater risk of various health problems, just as for humans. These include osteoarthritis, diabetes mellitus, pancreatitis, skin conditions, respiratory disorders, and urinary incontinence. And overweight dogs live less long than dogs of normal weight (amongst neutered male dogs, on average 5 months less for overweight German Shepherds, and 2.5 years less for overweight Yorkshire Terriers).[21] People whose dogs are overweight spend 17 percent more on health care for their dog and 25 percent more on medications.[22]

Beyond the physical effects, there are implications for the dog's happiness. One of the world's leading animal welfare scientists, Prof. David Mellor of Massey University in New Zealand, said to me, "We can anticipate that an animal that is just healthy, i.e., not seriously sick or even sick at all but is not fit, that its life experience is likely to be less enjoyable than an animal that is both healthy and fit."

A word on predatory behavior

Ghost was a picky eater. He also liked to find his own food, perhaps because he had been a stray and had had to do so to survive. When he stopped and cocked his head from side to side, listening in the long grass, I knew it was only a matter of time before he began to dig. The first time I watched with surprise as he dug up and ate a mouse. After that I knew he was listening to locate underground squeaks. Another time I was walking along with him on-leash when all of a sudden he stuck his head in a bush and came out with half a bird's nest and some baby birds in his mouth. Perhaps the worst time was when Ghost found a predeceased, bloated gray squirrel that he struggled to swallow, refused to swap for any kind of food, but eventually had to spit out because it was too big to go down in one gulp.

Predatory behavior is natural and methodical; it involves an area of the brain called the lateral hypothalamus and is part of what neuroscientist Jaak Panksepp termed the SEEKING system (see chapter 1).[23] The predatory sequence for wolves is orient → eye → stalk → chase → grab-bite → kill-bite → dissect → consume.[24] This sequence is preserved in some breeds like the Airedale Terrier but is modified in others.[25] If you've seen a Border Collie giving the eye to sheep, you've seen that the eye → stalk → chase part of the sequence is exaggerated, something they have been bred for. Meanwhile, the Anatolian Shepherd has been bred to round up livestock and the predatory sequence is inhibited. In everyday language, we may call predatory behavior aggressive, but unlike predatory behavior, aggression—whether offensive or defensive—is part of the RAGE system. In other words, predation and aggression involve different parts of the brain. With some dogs, cats and other small animals will always be at risk because the dog may perceive them as food, if not all the time, then perhaps when the small animal runs or makes high-pitched noises.

Our relationship with dogs and their food is not as simple as you might think. Some people need to pay more attention to what they feed their dog in order to ensure optimal canine health, while others arguably need to pay less attention in that every interaction with the dog should not be seen as a reason to feed. Perhaps most important is to pay more attention to our dog's weight. And to ensure they get enough water. Water should be available to dogs at all times; on outings, take a supply with you to offer when your dog might want it.

HOW TO APPLY THE SCIENCE AT HOME

- Choose the best diet for your dog that also works for you.

- Ensure good hygiene of both the place where you prepare food and the dog's bowls to reduce the risks of infection;

this is especially important if feeding a raw diet. Raw meat should be frozen first to prevent risks from parasites. If your household includes a child, a senior, or someone with a compromised immune system (including any of your pets), reconsider the feeding of a raw diet.

- Avoid too much sugar in the diet, as this is bad for the dog's teeth. Look out for healthy treats or use human food such as chicken in suitable amounts.

- Learn about overweight and obesity for dogs. If you're not sure whether or not your dog is a healthy weight, ask your vet. Reducing calories is the best way to get weight loss: weigh the dog's food to make sure you are giving the right amount. Treats (including those for training) should be calculated as part of your dog's daily calorie allowance.

- If most of your activities with your dog revolve around food, consider adding some new activities such as going for a walk, playing a game of tug, or petting your dog.

- Provide enrichment for your dog by varying their food and the manner of delivery (e.g., giving food toys).

12

SLEEPING DOGS

....

WHEN GHOST WAS sleeping, I used to marvel at how long he was. From nose tip to tail tip he was longer than I am tall. His breathing would slow right down. First a breath in with his big chest rising, then out with his chest going down, and then—pause—while I watched closely to check that another intake of breath would come. I loved the little quiet noises he made while dreaming, legs twitching away. Bodger makes similar wooflets and just occasionally he snores. Often he comes to sleep in my study while I'm working at the computer, sometimes lying so close behind my chair that I cannot move it, unless I want to wake him up and ask him to move.

These days, Bodger sleeps in a dog bed at the foot of our bed. Sometimes in the night I hear him get up and flop onto the floor, then later get back into his bed again. He had to earn the right to sleep in our room because it's the cats' domain and we needed to know we could trust him with them. I like that he sleeps in our bedroom, as it makes him feel like more of the family. I know some of my neighbors' dogs sleep in the house and others are free to roam outside, which means sometimes they run barking through our yard in the middle of the night. Clearly, people have different ideas about where their dogs should sleep.

WHAT SCIENCE SAYS
ABOUT CO-SLEEPING WITH DOGS

IN A STUDY of Labrador Retrievers, just over half (55%) slept indoors alone at night and 19 percent slept indoors with another animal.[1] Twenty-one percent slept indoors with a person, and for some of these dogs another pet was also present. Only 4 percent of Labs slept outside. In a representative study in Victoria, Australia, 33 percent of owned dogs slept on a dog bed inside, 20 percent on someone's bed, 24 percent in a kennel outside, and 3 percent outside without a kennel.[2]

It's an unfortunate tenet of some dog trainers that dogs should not be allowed to sleep on the bed or they will be "spoiled" and not behave very well. There is no evidence this spoils dogs, and "spoiling" dogs is not linked to behavior problems.[3] Of course, you may or may not want your dog to sleep on your bed, and that is entirely up to you. Some dogs, especially if they have body-handling issues, are at risk of being upset if disturbed and then growling or potentially biting. While reviewing the literature for this chapter, I came across a report of a dog who will growl or try to bite if the owner tries to make them get off the bed, and another who was allowed to sleep on the bed despite regularly biting the owner because the dog would bark all night long if shut out of the room. But many dogs will not be a safety risk to their owners or other pets and, I'm guessing, would like to cuddle up to their owners at night or at least be in the same room.

When dogs share their owner's bed or bedroom, it's called co-sleeping. Although it's not known exactly how common this practice is, one Australian survey found that 10 percent of people slept with their pet in or on the bed.[4] The researchers matched these people with others of the same age and gender who had completed the survey but did not co-sleep with pets. Although the people who let their pet sleep in the bed took longer to get to sleep compared with those that did not co-sleep with pets, the difference was quite small. And though they were more likely to say they were tired when they woke up in the morning, they did not sleep for any shorter or longer than those who did not sleep with their pets, and they did not report feeling any more tired during the day. The researchers concluded that, given how many co-sleep with pets, people must feel some kind of benefit from it.

A study by anthrozoologist Dr. Christy Hoffman of Canisius College, New York, asked women about who else sleeps in their bed—canine, feline, or human—and how their sleep was affected.[5] Perhaps surprisingly, dogs were the best sleep partner, perceived as providing more comfort and security than either another human or one or more cats. As well, women with a dog get up earlier, go to sleep earlier, and have more regular sleep schedules (something that helps with sleep) than those without. Amongst those who let their dog sleep on the bed, on average they thought the dog spent at least 75 percent of the night on the bed.

Sleeping in the same room as their owner may be what many dogs would prefer. In one study of dogs referred to a clinic for behavior issues, about 20 percent slept on their owner's bed, and of these, most were anxious dogs.[6] Other sleeping arrangements showed no difference between anxious and aggressive dogs. Of course, anxious dogs may prefer to be close to their owners, but for safety reasons people may choose not to allow potentially aggressive dogs to sleep on their bed.

Little research exists on where they prefer to sleep, but since dogs spend a lot of time resting and sleeping, dog beds must be important to them. A study of twelve laboratory Beagles found they preferred soft bedding.[7] When elderly laboratory Beagles were given a bed on the floor, they spent 83 percent of the ten-hour night-time there, compared with 21 percent of the time if the bed was 30 cm (12 inches) off the ground. Of course, elderly dogs may find a raised bed harder to get into, and it is not known if younger dogs would have the same preference. Another study of laboratory Beagles found they preferred to spend resting time in a bed, if it was available, rather than on the floor.[8] So it is important to give dogs a bed, and it seems a good idea to give dogs a choice of different types of bed and bedding.

"IT'S TOUGH TO narrow it down to one thing, but I'm going to go with enshrinement into law of basic Dog Rights, rather like the UN's Universal Declaration of Human Rights. This law would include basics such as freedom from frank abuse, pain and suffering, as well as shelter and soft bedding, freedom to move around, good food and medical care. But it would also include the right to engage in species-normal behaviors—sniffing, playing, chewing, interacting with other dogs if they wish, and corollary education documents so that people who keep dogs as companions or work with dogs recognize fear, worry, distress or other signs that a dog is not thriving. This would help people voluntarily comply. And if it doesn't count as a second thing, let's have vigorous enforcement worldwide."

—**JEAN DONALDSON,** author of *The Culture Clash: A Revolutionary New Way of Understanding the Relationship Between Humans and Domestic Dogs* and director of the Academy for Dog Trainers

UNDERSTANDING DOGS' SLEEP PATTERNS

SLEEP IS ESSENTIAL for good health, which is one of dogs' welfare needs. You may have noticed dogs sleep a lot. Young adult dogs sleep for between eleven and fourteen hours a day.[9] And just like people, dogs have phases of sleep, some in which there is rapid eye movement (REM) and some without. But there are differences in our sleep patterns too. One of them is obvious to all: dogs are often seen taking naps during the day. While most people sleep at night and are awake all day, dogs will sleep and then wake many times during a twenty-four-hour period. This is called a polyphasic sleep cycle, in contrast with our own monophasic pattern of a single block of sleep.

Dr. Deirdre Barrett is a Harvard psychologist and author of *The Committee of Sleep: How Artists, Scientists, and Athletes Use Their Dreams for Creative Problem-Solving—and How You Can Too* who studies sleep and dreams in people. She explained that our understanding of human sleep took a big leap forward in the mid-1950s when scientists used electroencephalograms (EEGs) to measure the pattern of electrical activity in the brain. They discovered the sleep cycle of REM versus non-REM sleep, Barrett said, that lasts "ninety minutes [in which] we alternate between very inactive brain sleep and then a stage when the brain is as active as when we're awake (though in different areas), which is rapid eye movement sleep, so named because the eyeballs move around like crazy. But EEGs were still expensive and rare at that time, so it wasn't until the 1960s that a lot of animal sleep research got done. Dogs along with a hundred other species were studied using the EEG hookup and found to have the same sort of sleep cycle varying between REM and non-REM sleep."

She went on to explain that the length of the sleep cycle varies depending on the species: "The brain is very quiet initially during sleep, and then without actually waking up the brain activity escalates a lot. That part's true for most mammals, including dogs, and there's a kind of size correlation with the cycle... Elephants go longer than ninety minutes and mice go way under ninety minutes." Indeed, a scientific paper from 1984 detailed the length of the sleep cycle of many different species—including the domestic duck, European hedgehog, sturgeon, and cockroaches. It reported that a study of six Pointer dogs found the length of a sleep cycle to be forty-five minutes.[10]

HOW MUCH SLEEP DO DOGS NEED?

OVERALL, DOGS ARE reported to sleep for 60 to 80 percent of the night and 30 to 37 percent of the daytime, although shelter dogs were found to have much less daytime sleep, perhaps because of the busy shelter schedule.[11] Low levels of light before bedtime seem to help dogs settle down to sleep, probably because the low light increases the production of a hormone called melatonin that helps regulate sleep. This is the same hormone associated with human sleep.

As dogs age, their sleep patterns change. A study of Beagles in three different age groups found young adults (1.5–4.5 years) are more active at night compared with older adults (7–9 years) and senior dogs (11–14 years).[12] During the daytime, senior dogs were less active than the other two groups. Interestingly, this study also found that feeding dogs twice a day rather than just once led to more activity at night, mainly because of an earlier start to activity in the morning. However, the effects of timing meals differently has not been studied.

A follow-up study looked at the amount of time dogs of different ages spent awake or asleep given a twelve-hour light–dark cycle.[13] At night, dogs were asleep for eight hours and seven minutes (younger adults) and nine hours and one minute (older adults), with seniors falling in between at eight hours and forty-five minutes total sleep time. During the daytime, young adult dogs were asleep for around three hours and nineteen minutes, older adults for three hours and fifty-nine minutes, and senior dogs for four hours and twelve minutes in total.

While it is normal for a dog's sleep pattern to change as they age, significant changes warrant a vet visit in case there is an underlying condition. Heart and thyroid conditions can affect sleep, and disturbances in a dog's sleep–wake cycle are one of the signs of canine cognitive dysfunction (CCD, similar to Alzheimer's disease in humans).[14] Dogs with CCD sleep more during the daytime and are awake and restless at night. This is thought to be due to changes in the circadian rhythm as a result of the disease process. (Other changes that may indicate CCD include the dog being more anxious, having less interest in interaction, and sometimes seeming disorientated.)

Puppies and senior dogs need more sleep than young adult dogs.
BAD MONKEY PHOTOGRAPHY

THE ALERTNESS OF DOGS

LAST NIGHT, I was woken from a deep sleep by Bodger leaping out of his bed and running around barking. I don't know what he barked at, since I was asleep, but he came back to bed and settled down again easily. A false-alarm bark, perhaps. I was so bleary-eyed I thought it must be nearly morning until I looked at my watch and saw it was just before 1 a.m. Fortunately this does not happen often, or I might want to banish him from the bedroom. What makes dogs wake up so suddenly when they are fast asleep?

It turns out dogs wake up fast in response to noise whether they are in an active or a passive stage of sleep. Australian researchers looked at the night-time sleeping and waking patterns of twelve dogs in their yards where they normally spent the night.[15] They called the sleep active or passive because they could not definitively say whether REM sleep was occurring or not. In the name of research, they played different sounds to see how the dogs would respond. The sounds included those that might be especially important to dogs (another dog barking once, and a sequence of repeated barks) or important to the owners (the sound of glass breaking and of rowdy young people discussing burglary, although of course the dog was not expected to understand the meaning of the conversation, just the rowdiness). As well, they played recordings of two noises that were not likely important to either dog or owner (a bus and a motorbike).

One of the conclusions will not surprise you: dogs responded more when they were alert rather than asleep. But, unlike humans who respond more to sounds during active (REM) sleep, the dogs responded at the same levels during active and passive sleep. Overall the dogs barked in response to 29 percent of the sound recordings. But—and this may not surprise you either—they were much more likely to bark in response to the sounds of barking. They also barked more if they were in a group of dogs rather than on their own. The researchers concluded this barking was likely to disturb people in the neighborhood, including the owners.

Dogs have many more sleep–wake cycles during the night than people do. Another study by the same Australian researchers observed twenty-four dogs at night-time.[16] Twenty of them were owned dogs (most of whom slept outdoors all night) and four lived at an animal house at the university. The scientists took video using red-light cameras and made observations of the dogs at night, including staking out the neighborhood from a nearby motor vehicle or an adjacent building. After fourteen months of both covert and overt observations, they calculated that during an eight-hour period at night, each dog had twenty-three sleep–wake cycles. On average, each cycle lasted twenty-one minutes, with sixteen minutes of sleep followed by five minutes in which the dog was awake. The dogs in fenced yards had a longer sleep time at nineteen minutes, while those who were free to roam typically had a sleep time of fourteen minutes (and were more likely to go out of range of the video camera).

One dog got no active sleep and had many sleep-wake cycles on her first night at the animal house, suggesting stress may have affected her sleep patterns. Another interesting finding was that when two dogs were sleeping together, their sleep-wake cycles did not synchronize and were not the same—except when both were woken at the same time by another dog barking.

These detailed observations show dogs have many short sleep-wake cycles through the night. They also show that dogs who sleep outside may engage in activities other than sleep and be disturbed by other dogs and people in the neighborhood. Dogs may have fewer disturbances if they sleep inside.

THE ROLE OF SLEEP

WE ALL KNOW the feeling when something bad happens during the day and then we just can't sleep at night. It turns out that, like humans, dogs' sleep is affected by bad experiences—but the effects are not quite the same. In a study of sixteen dogs published in *Proceedings of the Royal Society B*, scientists used an EEG to measure the effects of stress on dogs while they were sleeping after a good or bad experience.[17]

Over the course of three sessions, of which the first was a practice, the dogs had a six-minute-long good or bad experience, followed by three hours of sleep. In the good experience, the dog was petted every time they went to their owner, was spoken to nicely, and played fetch or tug as they preferred. In the bad experience, the dog's leash was tied to the wall and they were left alone for two minutes, then their owner appeared but ignored them, and last an experimenter approached in a threatening manner and sat looking at the dog without responding to them.

During the three hours after the bad experience, the dogs got an average of seventy-two minutes of sleep and the duration of a sleep cycle was fifty-six minutes. After the good experience, the dogs took longer to go to sleep, and on average they got sixty-five minutes of sleep with a sleep cycle of fifty-one minutes. After the negative experience, dogs had a longer period of REM sleep, which was expected because REM sleep is associated with emotional processing. Non-REM sleep, which is when the deepest sleep occurs, was higher after the positive experiences. After negative experiences, the dogs got less deep sleep. These changes in sleep after stress are thought to be a protective response.

The researchers also found the dogs' personalities were linked to how they behaved with the owner. For example, dogs who were rated as more agreeable and less open hid behind their owner more when the experimenter was sitting and looking at them in the negative experience. In turn, some of these behavior differences were linked to changes in the sleep cycle. What this means is that individual differences in how the dogs responded to the experiences were also reflected in changes in their sleep. Although we already know that human sleep is affected by stressful events, this is the first time good or bad experiences have been shown to affect how well a dog sleeps.

THE RELATIONSHIP BETWEEN SLEEP AND LEARNING

ANOTHER STUDY FOUND that learning influences sleep and sleep influences learning.[18] Fifteen pet dogs took part in both a learning and a non-learning condition on different days. In the non-learning condition, the dogs practiced two commands they already knew in Hungarian—sit and lie down—prior to the sleep session. In the learning condition, the dogs were taught the English-language commands for these two behaviors. After each session, the dogs were given time to sleep, and recordings were taken of their brain-wave activity (the technical term is polysomnography). Results showed changes in their brain activity during both non-REM and REM sleep after learning; these changes are consistent with the consolidation of memories during this time. After the three-hour sleep session, the dogs' performance at the new-language commands was improved.

In a second experiment, fifty-three pet dogs were given the same task of learning the English commands "sit" and "lie down," which they already knew in Hungarian, and were then assigned to four different one-hour activities: sleeping, going for a walk on-leash, doing another learning activity, or playing with a food-stuffed Kong toy. They were then tested on the new commands immediately and again a week later. At the immediate re-test, the dogs who had slept or been on an on-leash walk did best; it was thought the play had excited the dogs and the other learning task had interfered with the consolidation of memories. At the one-week re-test, the dogs in the sleep, leash walk, and play conditions all performed well, but the dogs in the other learning condition still performed badly. These results suggest that what dogs do after learning matters; giving the dog a mental break seems to help over the long term, whereas asking the dog to do another learning activity hindered their learning. The dogs' regular sleep at home during the week after the learning session seemed to help with consolidating memory.

The role of sleep could explain the results of another study which looked at how often dogs should be trained.[19] Scientists trained forty-four laboratory Beagles to go to a basket and stay there, a complicated task that was broken down into eighteen steps. Half of the dogs were trained once or twice a week and half were trained daily; in addition, some of the dogs had short training sessions and some had longer training sessions. The results found that the dogs taught once or twice a week performed better than those taught every day (although obviously it took longer for them to have enough training sessions to learn the task). It was also found to be better to have a shorter training session. However, when dogs were tested four weeks later, they all remembered and went to their basket on command. The researchers suggested that the shorter sessions required more cognitive effort (and so led to better memory) and that more opportunities for sleep between sessions also helped with memory. Taken together, these studies suggest that letting dogs get a good night's sleep between training sessions will help with learning.

WHY ANIMALS TWITCH
AND WHIMPER WHILE SLEEPING

WHEN I SEE Bodger's legs twitching in his sleep, I like to think he is dreaming of running after a ball. In Ghost's case, I would say he was dreaming of chasing rabbits. I asked Dr. Deirdre Barrett if we can infer from this movement that they are dreaming. "No, for sure," she said, and compared sleeping dogs with sleeping humans. "Most dreams are associated with rapid eye movement sleep." During REM sleep, the muscles are temporarily paralyzed so that we can't move around. Barrett explained that sleepwalking, when people get up and move around during their sleep, happens in non-REM sleep and so there is not an associated dream. "I think that in dogs the best guess is that the majority of their movement is probably not accompanied by a dream, it is just that their motor area is suddenly a little more active and that there's not a lot of content with it," she said.

And as for those little noises they make in their sleep? Barrett explained, "The same is true for sleep talking. About 80 percent occurs out of non-REM sleep of humans and does not have a dream associated with it, and about 20 percent of human sleep talking is straight out of dreams and is associated with dream content. So, again, those studies have not been done in dogs." I was disappointed to discover that twitching and whimpering didn't mean my dogs were dreaming, but now I know! And it didn't stop me from wondering what dogs dream about.

"Obviously we can't interview them," said Barrett. "Human studies are done by asking people what they're dreaming about, I mean sometimes while looking at their EEGs and other things at the same time, but always by getting a dream report . . . It's sort of borderline with gorillas. Penny Patterson says that Koko the gorilla signs what she thinks or dreams when Koko wakes up in the morning. And I think that's plausible, although certainly not proven. But you know, with dogs we're never going to have that."

However, she makes some connections with what we know about humans. "So with humans, humans dream about the same things that they are most concerned with by day: the people important to them, problems facing them, environments that they're frequently in. Even though you see those transformed in kind of distorted fanciful ways, the basic content is the same content as their waking thought even though it's in this kind of less logical linear and much more visual emphasis way. So I think it's safe to assume that dogs who are pets must dream about their owners a lot. I mean humans dream about their significant others, and dogs are so focused on us by day that I think they must surely dream about us at night. And probably their favorite meals and toys and the parks they run in are getting mashed up in some combination in their dreams."

Maybe Ghost did dream of chasing rabbits after all—just not when his legs were twitching.

HOW TO APPLY THE SCIENCE AT HOME

- Know that puppies need more sleep than adult dogs. Changes in sleep patterns happen with age, just like in people, but any sudden or significant change warrants a visit to the vet in case of an underlying condition.

- Give your dog happy experiences during the daytime to help them get a good night's sleep.

- Choose whether or not to let your dog sleep in your bedroom or on your bed, and be consistent.

- Make sure your dog has comfy beds to sleep in.

- If you have a busy household with lots of people at home all day, ensure your dog gets some quiet time to sleep during the daytime.

- Make sure your dog gets a good night's sleep after a training session to help consolidate memory and improve learning.

- If you have to go to a new location, your dog might not sleep as well on the first night, so bring some bedding from home to make them feel more at ease.

13

FEAR AND OTHER PROBLEMS

....

WHEN GHOST ARRIVED to live with us, he would not eat. We took him to see the vet, who went to find a can of cheap dog food. As he put it into a bowl, a large blob dropped onto the floor. Ghost ate it. The vet offered the bowl to Ghost, but he turned his head away and licked his lips, both signs of stress. Then the vet deliberately dropped some food on the floor. Ghost ate it right away. "He's afraid of the food bowl!" said the vet. In fact, it seemed to be any noise made by the bowl that he did not like. If I took care to put it down very quietly, he would approach it, and before long he was fine with it.

Another fear took longer to go away: Ghost would shy away from our hands if we reached towards him. I learned to put my hand out and wait for him to decide if he wanted to approach or not; given the choice, he often would. He loved to be petted, but on his terms. It took a whole six months before he stopped flinching away from our hands as if they would hurt him.

Unfortunately, fear can take a long time to resolve, if at all, but is easy to instill. Fear is behind many behavior problems, but medical issues, boredom, lack of physical exercise, lack of opportunities to use the brain, not being provided with needed resources (like chew toys), and simply not knowing the rules can all contribute to behavior problems too. The most common behavior problems reported by owners are aggression and fear/anxiety, including fear of noises and separation anxiety.[1]

Owners' perceptions of problem behavior matter, not just because some issues may not be noticed (as with some cases of fear), but also because what constitutes a problem is in the eyes of the owner. Some people may tolerate the occasional house-training indiscretion while for others it will be a serious transgression. Behavior issues can be a welfare issue for the dog and problematic for the owner, but more worrisome is if owners do not recognize welfare issues, because it means they will not address them.

Undesirable behaviors are considered one of the most important issues for the welfare of dogs, according to a survey of experts (the other important issues being inappropriate husbandry and lack of owner knowledge—both of which, of course, may contribute to undesirable behaviors).[2] Noise phobias and separation-related behaviors are highlighted as a particular welfare problem.

"IF I GOT to choose just one thing that would make the world better for dogs, I'd ask humans to take their dogs' misbehaviors less personally. Your dog isn't trying to be boss or ruin your day; he's trying to get his needs met. Imagine if each training challenge we faced—to create a more responsive, cooperative, civilized dog—was viewed as our invitation to become more curious about how behavior functions and changes. Then every lunging, barking or biting dog would be recognized as a learner, evolutionarily prepared to adapt his behavioral repertoire to the changing environment. What freedom we'd have to stop punishing and to start exploring the joy of true dialogue with another species."

—KATHY SDAO, MA, applied animal behaviorist, Bright Spot Dog Training, and author of *Plenty in Life Is Free: Reflections on Dogs, Training, and Finding Grace*

Another welfare issue occurs when people mistake their dog's behavior for guilt, spite, or stubbornness. There is a certain look with a bowed head and pleading eyes that many people believe means the dog is feeling guilty. But in order to feel guilt, dogs need to know they have done something wrong (otherwise there is nothing to feel guilty about). It's important to note we cannot definitively say whether or not dogs sometimes feel guilt, but we do know that the look so many of us perceive as guilt is not linked to wrongdoing. This was tested by Barnard College dog cognition scientist Dr. Alexandra Horowitz, who invited fourteen owners and their dogs to come to the lab.[3]

A cookie was placed on the table, and owners were asked to tell their dog not to eat it; then the owner left the room. Over four repeats of the scenario, sometimes the dog got to eat the cookie and sometimes they didn't. When the owner returned to the room, they were told if their dog had eaten the cookie or not— but the information was true only half the time. Owners scolded the dog when they believed the cookie had been eaten—and video analysis showed those canine behaviors we associate with the "guilty look" were not related to whether or not the dog had done something wrong but to whether or not the owner scolded the dog. Interestingly, the so-called guilty looks were most pronounced when the dog had not eaten the cookie but the owner scolded them believing they had.

Some people say their dog looks guilty even before they have discovered the thing the dog did wrong. Animal researcher and *Dog Spies* blogger Julie Hecht and colleagues devised an experiment in which pet dogs were left alone in a room with a piece of hot dog they were not meant to eat.[4] A questionnaire completed by the dog owners showed most of them think dogs can feel guilt, and they said they scold their dog less because of that guilty behavior. But there was no difference in the "guilty behaviors" between dogs who had eaten the food and those who had not. Although individual dogs showed subtle differences in their greeting behavior in the different scenarios, owners were not reliably able to detect whether or not their dog had transgressed and eaten the food.

A better understanding of dog behavior would go a long way to helping people prevent and deal with behavior problems. The rest of this chapter will consider specific behavior issues and what can be done to help.

FEAR

SOMETIMES AT NIGHT I hear the call of the barred owl, *who-cooks-for-you, who-cooks-for-you-all*. I used to like it until Bodger came to live with us. Bodger is terrified of this noise. He erupts into a fit of barking and takes a long time to calm down afterwards. I understand his fear because the barred owl is huge. Could it take a small dog? I don't know, but I am pretty sure it could not take a dog the size and weight of Bodger (he is on the big side for an Australian Shepherd). But in any case, he is safe and sound inside the house. Sometimes I sit with him in the early hours, petting him (because he wants it and it seems to help) and showering him with treats when the owl calls. It helps in the moment but leaves me tired the next day.

I decided to do something about this fear and found a recording of the barred owl on the internet. I played it in my study one day when Bodger was in another room, setting the volume very low in the hope he would be okay with the sound at that level. But there was an instant eruption of growls and woofs. Bodger came running into the room barking and looking for the owl. My attempt at desensitization had gone completely wrong; it is only desensitization if the dog is relaxed and happy. I tried to rescue the situation by stopping the recording right away and giving Bodger some of his favorite sausage treats, which is counter-conditioning without desensitization (chapter 3). The treats were for simply having heard the owl, and he would have received them no matter what behavior he displayed. I've done this several times now, and getting Bodger used to the owl is a work in progress. Meanwhile, I sometimes wish the owl would move along to someone else's yard.

Bodger's reaction to the owl was one of fear. Fear causes a physiological response and activates what we call the fight-or-flight reflex, because an animal's response to imminent danger is typically fight, flight, fiddle, or freeze. In a dangerous situation, this primitive response can help us survive. Although anxiety looks very similar to us when we are looking at a dog, it is not the same as fear. Anxiety is a longer-lasting concern about a possible threat. Since the outward signs are the same, and we can't say if the dog is responding to something imminent or future, it makes sense to think of fear and anxiety as a spectrum.

When a dog is fearful, people are often quick to blame something that must have happened in the dog's life. In fact, there are several potential causes of fear. Lack of socialization, genetics, prenatal stress of the mother, early life experiences, and bad experiences at any time of life can all cause fear. Simply not having enough positive experiences during the sensitive period for socialization, from 3 until 12 to 14 weeks, is enough to cause fear of novel things or people (see chapter 2). In a study of guide dog puppies, those who had a frightening experience with another person during the sensitive period were more likely to be afraid of people as an adult dog.[5] Similarly, those who had been threatened by another dog during the sensitive period were more likely to be fearful of other dogs as an adult.

To some extent, fear and anxiety are genetic. While we used to talk about behaviors being either nature (biological) or nurture (cultural), it's become increasingly clear how closely they are intertwined. Some studies have even identified specific genes associated with fearful responses in particular breeds of dog.[6] Good socialization is essential, but it is not the only part of the puzzle. If you can, try to see both of the dog's parents before you get a puppy—and check that they are friendly, not fearful, dogs.

Some of the causes of fear start even before the puppy is born. This phenomenon is not specific to dogs, and research on species such as humans, cats, and rats that similarly have to take care of their young are relevant here. Prenatal stress may affect a puppy's development as the mother produces stress hormones that will circulate to the fetus. These stressful experiences can cause epigenetic changes, which can affect gene expression, according to ongoing research.[7] One such mechanism is a chemical process called the methylation of DNA that can basically turn the gene off. Another process is histone modification, which affects how easy it is for parts of the gene to be transcribed. As well, processes involving RNA can also affect gene expression. These epigenetic changes affect the whole system, not just the brain, and they can be adaptive if they help the animal fit with the environment, or maladaptive if they are not needed for the environment the animal will live in.

Stressful experiences can also affect the development of the nervous system and hormones in the fetus, and this may continue into the period after being born. Prenatal stress can affect the behavior of the young; when mother cats are subject to stress from not having enough to eat, their kittens show abnormalities of behavior development, including poor balance and fewer social interactions with the mother.[8]

Maternal care of puppies during their first few weeks can also influence stress and anxiety in adult dogs. Because puppies are born deaf and blind, the mother's behavior towards them in those early weeks is very important. As well as nursing the puppies, the mother licks their anogenital region to stimulate urination and defecation, protects the puppies from harm, provides body heat while they are too young to regulate their own body temperature, and interacts with them through nuzzling and licking. The puppy may play with others in the litter and start to explore the environment. One study of twenty-two litters of German Shepherd puppies being raised for the Swedish military found that better maternal care—as measured by time in the box, in contact with the puppies and nursing them—was linked to more social and physical engagement when the puppies were adult dogs.[9] Another study found that puppies receiving more maternal care during the first three weeks of life were already more exploratory and less stressed at 8 weeks of age.[10]

Of course, traumatic experiences can also cause fear. A questionnaire study published in *Frontiers in Veterinary Science* found many people attribute behavior problems in their dog to a traumatic experience, especially when the dog has experienced two or more such events.[11] Forty-three percent of the dogs had experienced at least one of the following traumatic events: change in owner, time spent at a shelter, being lost for more than a day, changes in the family (such as the birth of a child or people moving out), traumatic injury, long-term disease, or surgery. These dogs were more likely to be unhealthy, showing the effects of long-term stress on health. Owners can mitigate some of these factors; for example, teaching a good recall and ensuring the dog has identification so they are not lost, but these results also show the importance of helping dogs to cope with trauma.

In chapter 5, I wrote about how some people find it hard to recognize the signs of their dog's fear at the vet. A study in PLOS ONE found that actual experience with dogs, such as working as a trainer or groomer for many years, helps people to recognize the signs of fear.[12] One reason may be that they pay attention to more parts of the dog, so they are likely to see and put together the different signs. In particular in this study, they paid attention to the ears, which less experienced people did not. The eyes, ears, mouth, and tongue were all found to be useful in spotting fear. Unfortunately, simply owning a dog did not count as experienced in this study.

Just like trips to the vet, we know that fireworks and other loud bangs can be terrifying for dogs. One large study asked people if their dog was afraid of noises, and 25 percent said yes.[13] However, when a subset of people took part in a structured interview in which they were asked about specific loud noises, such as the vacuum cleaner and fireworks, and specific behavioral responses, such as trembling or seeking out people, 49 percent of the owners reported that their dog demonstrated a fearful response around loud noises. Forty-three percent said they had seen their dog trembling or shaking in response to a noise, 38 percent reported their dog barking, and 35 percent reported their dog seeking out people. The reality is that many people could seek help for their dog's fear of loud noises but do not because they don't realize there is a problem.

If you have a fearful dog, your priority is to help your dog feel safe. Forcing your dog to face their fears will likely make things worse. If you have been using aversive methods to train your dog, stop, because this adds to your dog's stress. It is fine to comfort them if you think it would help (not all dogs want it), but you should also work out a strategy to help in the longer term. This may involve behavior modification and seeing your veterinarian to determine if medication will help.

AGGRESSION

AGGRESSION DOES NOT just mean biting. A survey published in *Applied Animal Behaviour Science* defined aggression as "barking, lunging, growling, or biting."[14] Three percent of dog owners in this survey said their dog is aggressive towards family members, 7 percent reported aggression to people they don't know when they come into the house, and 5 percent said their dog is aggressive to people they don't know when out and about. The same dogs were not in all three categories. For example, a dog that was aggressive to family members was generally not aggressive to unfamiliar people, whether in the home or outside it. This means that if a dog is aggressive in one situation, it may not be aggressive in a different one. The thing to remember is that any dog can bite if put in a situation where they feel threatened.

Unfortunately, some governments in Canada, the US, and the UK have introduced breed-specific legislation (BSL) that regulates or bans owners from having certain breeds of dog in order to decrease dog attacks on humans or other animals. For example, Ontario bans pit bulls, which it defines as Staffordshire Bull Terriers, American Staffordshire Terriers, American Pit Bull Terriers, and any dog with "an appearance and physical characteristics that are substantially similar."[15] In some parts of the US, breeds including Chow Chows, German Shepherds, Rottweilers, Doberman Pinschers, and American Staffordshire Bull Terriers have been banned.[16] However, this legislation mistakenly gives the impression that some breeds are "dangerous" and others are "safe."

In the UK, which banned four breeds (Japanese Tosa, Doga Argentino, Fila Brasileiro, and the pit bull terrier) in 1991, dog bites have continued to rise, causing organizations including the RSPCA to campaign for an end to BSL.[17] In Ireland, where hospitalizations due to dog bites have also increased, there is concern that BSL is contributing to the risks.[18] A study of dog bite injuries in Odense, Denmark, found that banning certain breeds had no effect, and nor did requiring them to be muzzled and on-leash in public; in addition, most injuries occurred in private (not public) spaces.[19] There is no evidence that BSL works; instead, it penalizes people who have well-behaved dogs who happen to be amongst the banned breeds. One study even found differences in what is thought to be a "pit bull" in the UK and the US.[20] As a result of difficulties in identifying breeds, accurate information about the breeds of dog responsible for bites is not available.

In contrast to BSL, in 2000 the city of Calgary, Alberta, introduced a model based on responsible ownership that has proven effective at reducing the number of dog bites.[21] This is now known as the Calgary model. Instead of banning particular breeds, the city has focused on public education campaigns about dog safety and on strong licensing and bylaw enforcement, emphasizing that owners should obtain their dogs from an ethical source, get them spayed or neutered, socialize and train them, and not let them run at large or be a nuisance. Dog bites and aggressive incidents involving either dogs or cats must be reported to the city, and when there is a complaint, bylaw officers can help the owner of an aggressive dog work out a solution that may include training. The end result is that while Calgary's population has grown substantially, the number of aggressive dog incidents has fallen from more than 2,000 in 1985 to 641 in 2014, of which 244 were bites.[22]

One of the problems with trying to prevent dog bites is that people may not realize they are at risk. Over the years, people have begun to believe they should be able to do things like take food away from a dog, which in the past would have been seen as pretty stupid. When scientists interviewed people who had been bitten by a dog, one of the things they found was a belief that "it won't happen to me" and that people trust their own dog.[23] One thing that does correlate with a dog biting someone is the dog having shown previous signs of aggression, such as lunging, barking, or growling at people.[24] So if you see this behavior in your dog, seek help sooner rather than later. Studies also show a correlation between the use of aversive dog training methods and aggression (one of several good reasons to use positive reinforcement instead).[25]

SEPARATION ANXIETY

DOGS WITH SEPARATION anxiety become very upset when they are apart from their owner. Signs include destruction, such as chewing and scratching, often close to the exit from which the owner left; vocalizations such as whining, barking, and howling; and possibly urination and/or defecation as a result of the stress.[26] Importantly, these signs only occur when the owner is absent and they happen within minutes of the owner leaving, although the dog may show signs of anxiety when the owner is preparing to go out (putting shoes on, etc.). Destructive behavior while the owner is out can have other causes, such as boredom or something scary happening during that time, so video of your pet may help in making a diagnosis.

The good news is that separation anxiety is treatable, though it takes time and effort. In a clinical trial, dogs with separation anxiety underwent a standardized twelve-week program that started with short absences from the owner and progressed systematically to longer absences.[27] No medication was included. The dogs in this program showed improvement in their anxiety compared with dogs who received no treatment and did not improve at all. It seems likely that a program tailored to the individual dog would be even more beneficial, and this is how separation anxiety treatments work.[28] Treatment also often includes the use of medication prescribed by a vet or veterinary behaviorist.

RESOURCE GUARDING

RESOURCE GUARDING MEANS protecting (or trying to protect) resources such as food, food bowls, or dog beds from other dogs or people. It is often seen as being due to a fear of resources being taken away, perhaps an evolutionary hangover from when resources may have been scarce. Resource guarding is associated with rapidly ingesting food, using strategies to avoid having a resource taken away (e.g., by moving it to another place or putting their body in the way), and demonstrating threats and aggression. One study published in *Applied Animal Behaviour Science* found people who looked at videos of resource guarding were good at recognizing it when signs of overt aggression (biting and snapping) were involved, but not so good at spotting the earlier signs.[29] Threats such as freezing or growling, baring the teeth, and tensing the body were spotted more often than when dogs ate rapidly or avoided having the resource taken away. If people learned to recognize all these signs, they could take action before the dog becomes aggressive. People who went to dog training classes or who said they had experience in dog behavior were much better at recognizing those early signs, suggesting education is beneficial.

Research has shown that when resource guarding is identified in dogs in shelters, it does not necessarily occur in the new home; similarly, dogs who did not guard in the shelters may do so at home.[30] As a result, many shelters no longer test for resource guarding. Taking the dog's bowl away during meals is linked to more serious and more frequent guarding, so don't do that.[31] In contrast, adding nice food to the dish during mealtimes is linked with less-serious guarding (but may not be safe to try with serious cases). Resolving resource guarding may involve teaching the dog it is okay for the resource to be approached and taken away (desensitization and counter-conditioning), teaching them to drop items, or training a behavior such as "sit" that is impossible to do at the same time as guarding (technically known as differential reinforcement of an incompatible behavior). Because it is so easy to miss signs (such as a freeze), it would be advisable to seek professional help.

HOUSE-TRAINING ISSUES

HOUSE-TRAINING ISSUES CAN be a common problem, as dogs must be taught not to soil the house. House training involves taking the dog outside to toilet often enough that they do not get a chance to go in the house; young puppies cannot hold their bladder for long and may need to be taken outside every thirty minutes. As well, it is important to accompany the dog and reward them as soon as the peeing or pooping is finished. This means planning ahead so you always have treats handy. In case of accidents in the house, use an enzyme cleaner to completely remove the smell, because dogs' noses are much better than ours. Always do a patch test with the cleaner to check it is safe to use on your flooring.

Unfortunately, sometimes people yell at or punish their dog for accidents in the house. This is essentially the opposite of house training. What can happen is that your dog becomes afraid to toilet in front of you and instead will wait until they are home and you are not there. Medical problems can be a common cause of house-training issues, so any time a previously house-trained dog has an accident in the house, see your veterinarian. Small dogs are said to be harder to house-train, but the reasons are not understood.[32]

IT'S WORTH GETTING HELP

WHEN YOUR DOG has a behavior problem, it can feel very difficult. It's important to remember to take care of yourself, as this helps us be more resilient. As well, remember that your dog's behavior is not due to spite or stubbornness, and try to feel some empathy for your dog. Although there are some good resources in the form of books and websites, there is also a lot of poor information. If you are hiring a dog trainer to help, look for someone with a qualification (see chapter 3), who is a member of a professional organization, who will use reward-based methods to train your dog (not aversive methods that risk making things worse), and who engages in continuing professional development. This information should be on their website. In some cases, your vet may refer you to a veterinary behaviorist, a veterinarian specially trained to help with behavior problems who can prescribe psychotropic medications.

Some behavior problems are due to an underlying medical condition. It is hard for dogs to tell us when they don't feel well, but any sudden change in behavior, such as a dog who is already house-trained starting to urinate in the house, warrants a visit to the vet to rule out medical issues. Some recent research even suggests that pain may be behind late-onset noise phobias.[33] Resolving medical issues can be part of or even the entire solution to some common problems.

The good news is that the right help can make a real difference to all kinds of behavior problems. In a study of dogs referred to a veterinary behaviorist for problems including anxiety, resource guarding, and aggression related to fear, the results were very positive.[34] While 36 percent of owners had thought about either rehoming or euthanizing their dog before their first consultation, only 5 percent had done so three months after the consultation. The results showed a number of risk factors, other than the dog's behavior problem, are linked with rehoming/euthanasia decisions, in particular factors to do with the owner and environment.

"If you look at the data," said veterinary behaviorist Dr. Carlo Siracusa, one of the authors of the study, "it tells you that having experienced a traumatic separation [is a risk factor]. So for example having lost a dog, or having a kid that departed to go to college, or having experienced the loss of a dear person, this might put a psychological pressure on you. When you are grieving from a loss or a separation, you are more vulnerable to possible problems." He explained, "If you get a new dog and the dog is in good health, doesn't have a behavior problem, you might just be able to enjoy his company and everything goes well. But if soon after you get a dog, because you are in a vulnerable state and probably you were relying on the dog to help you go through your grieving, to help you heal, and then the dog himself turns out to have a problem, then people's ability to tolerate the problem is much lower, because their compassion is exhausted; they came from a traumatic experience."

Siracusa says a dog's behavior problems can affect your relationship with the dog. "Many times the behavior problem makes the dog very different from what you would have liked it to be," he said. "We have people telling us, 'This is just not the relationship that I was thinking to have with the dog.' If there is a problem of aggression, for example, and especially aggression towards the family, being unable to medicate the dog, being unable to give care to a dog, these are risk factors for rejection." He added that even if veterinary behaviorists explain that the dog is feeling threatened and is anxious, some owners say, "'Yes, I do understand but this is not what I wanted from a relationship with the dog.'" And for them this might be a reason not to address the behavior or to give up on the dog. Instead, Siracusa recommends that if home training is not working to resolve a problem, people should see their vet who might recommend a veterinary behaviorist.

He says help may involve medication for the dog, education on dog behavior, and learning to read canine body language and de-escalate situations earlier. His final piece of advice is for when things are very difficult: "Before making a choice, especially if it's a choice that owners make when they are under stress, for example if your dog bites someone, don't make the choice to put your dog to sleep or surrender your dog when you are still very affected by the episode. You need to calm down and think of it a bit more rationally; talk to your veterinarian, see a veterinary behaviorist [vb] if it's necessary. Again, the vb might not be able to fix your dog but he can help you to manage the problem a lot." Ultimately what this study shows us is that treatment for a dog's behavior problem can prevent rehoming and save dogs' lives.

Behavior problems are not just a concern in and of themselves. Fear, anxiety, and medical issues can also limit a dog's ability to learn or take advantage of enrichment opportunities presented to them. A sense of safety and control is important for dogs. Treatment for behavior problems will almost always include helping the dog to feel safe, including stopping any use of punishment.

Over the years, I have worked with many fearful shelter dogs, throwing treats in their direction from a safe distance, making myself as small as possible, and avoiding eye contact, waiting to see if they will choose to approach me. The dog always has a choice whether to stay in the kennel with me, hide under their bed, or exit via the dog door and wait outside, from where they sometimes come to peer at me and stick their head through to eat a piece of chicken. When finally one of these dogs chooses to come and lie down next to me, I'm thrilled, but I don't move. My aim is always to help the dog feel safe.

HOW TO APPLY THE SCIENCE AT HOME

- If you have a pregnant dog, try to keep her routine predictable and reduce stress so that she feels safe. This is good for the behavioral health of her puppies.

- Learn the signs of fear, anxiety, and stress so you can recognize them in your dog (e.g., tucked tail, low body posture, shaking and trembling, seeking out people, hiding or withdrawing, etc.).

- Discuss any sudden change in behavior with your vet in case a medical issue is the cause.

- If your dog has a behavior problem, don't use punishment. It does not teach your dog what to do instead of the problem behavior, and it interferes with the dog's feeling of safety.

- Make it a priority to help your dog feel safe. A fearful dog is not able to take advantage of positive experiences offered to them.

- If your dog has a behavior problem, know that you are not alone. Seek help sooner rather than later as this can stop problems from getting worse and lead to an earlier resolution. Good behavior advice and treatment can make a huge difference.

14

SENIORS AND DOGS WITH SPECIAL NEEDS

. . . .

BODGER COUNTS AS old now. The years have gone by in a flash. Why can't he still be young?! But he is still happy and bouncy. And he still has the herding dog's desire to keep an eye on everyone and know where they are, preferring it when we are all together. I think he will always be excitable, although he has learned to sit nicely and wait for a treat when something potentially exciting occurs, like a bicycle coming up the road or a person carrying a garden hoe. Sometimes I hop or dance just to excite him. I like to see his facial expression change, his eyebrows raise, and his ears prick up. Then suddenly he gets up and runs to grab his rope. It's hard to imagine him slowing down. Ghost, on the other hand, began to slow down far too soon, the health problems he'd had since the beginning not helping. When he began to struggle to get in and out of the car, we found a step he could use. His walks were slower, shorter, and not always with Bodger who wanted to race ahead while Ghost took his time. His routine was the same, and since we are almost always home, he still almost always had company. His size was not in his favor. Large dogs just seem to age sooner.

We can't say a human year is equal to X dog years because puppies develop rapidly and so these years, perhaps equivalent to childhood, count as a lot, whereas the rate of maturation varies throughout the lifespan. Different breeds have different lifespans, and larger breeds do not live as long, so one way to think of it is to group by breed size. Using this kind of measure, large and giant breeds—dogs who weigh over 22.7 kg (50 pounds)—count as seniors when they are between 6 and 8 years old, and are considered geriatric once they reach 9 years. It's a little better for medium and small breeds, which count as seniors once they reach 7 and are considered geriatric once they reach 11 years. So they become senior later and stay seniors for longer. In a study of Border Collies, they were classified as late puppies from 6 months to a year; adolescents from 1 to 2 years; adults in three stages— early (2–3 years), middle-aged (3–5 years), and late (6–8 years); seniors from 8 to 10 years; and geriatric at 10 years or over.[1] At 10 years old, Bodger is a senior.

WHAT HAPPENS TO DOGS AS THEY AGE?

IN MANY WAYS, the changes in dogs as they age are similar to those we see in humans. Since this makes dogs a useful model for some kinds of human aging, we are likely to see more research into dogs as they age. Aging affects all of the dog's systems.[2] Some of the changes are visible to us: a coat that's less bright than it used to be, less muscle, some changes to sleep patterns, and changes in how often they come to interact with us. They are likely to be less active, less playful, and less excited about walks, although some dogs continue to be athletic as they age. Most dogs will have changes in their body composition, with a higher proportion of body fat, although this is less so in dogs that remain active. This reduction in activity and changes to the metabolic process mean older dogs need about 20 percent less energy intake, unless they are still very active. But while they need less food, they have a higher requirement for protein—50 percent more according to one study. They can have changes in eating and drinking habits such as not finishing a meal, drinking less, spending less time chewing, and finding it more difficult to locate food dropped on the floor.

Older dogs may find it harder to jump onto furniture or get into the car. Changes in the hair follicles may lead to the appearance of white hairs, especially on the muzzle and face. Older dogs gradually lose their night vision, and changes to the nucleus of the lens in the eye make it more dense and cause it to develop a bluish haze (although this usually does not affect vision). Some dogs will have cognitive changes, while others will not. Some changes may take a while for us to pick up on, as with gradually reducing vision or loss of hearing. Vision problems may not be obvious unless hearing is affected. Other changes are not visible until the vet tells us.

Older dogs are at greater risk of cancer, cardiovascular problems, kidney problems, periodontal disease (especially in smaller dogs), and diabetes mellitus. Aging also affects the endocrine system, including the adrenal glands, which means older dogs do not cope as well with stress. For us as owners, it is hard to know what is due simply to aging healthily and what we should discuss with the vet. The vet may ask to see the dog more often, perhaps twice a year for a health check instead of once, to pick up on any problems sooner rather than later.

It's important to help elderly dogs remain part of the family and continue to have the kinds of experiences they enjoy. In an article in the *Veterinary Record,* Dr. Naomi Harvey of the University of Nottingham wrote of how the fear of one day losing her cat, and knowing that day was getting closer, inadvertently led her to spend less time with and pay less attention to the cat.[3] But the cat still needed her usual fuss, attention, and play, and once Harvey realized this, she could make a point of spending more time with the cat again. She told me, "Elderly dogs may have different needs but can still be wonderful companions and deserve the same attention and TLC as younger, more active dogs."

She added, "It's a sad fact that many people would prefer to rehome a puppy over an older dog, and studies have shown that people often become less attached to their dogs as they age. Caring for an elderly pet means adjusting, but small adjustments can make a big difference. They may not be able to walk, run, or tug any longer but they still have active minds and enjoy the stimulation of sniffing and smelling things, and you can interact with them through training and brain games. If they can't walk far, don't leave them behind; consider taking them out on a pet trailer, stroller, or carrier so they still get the experience of going with you and seeing/sniffing the sights and smells with you."

As dogs get older, they experience a decline in attention just as older people do. One study took pet dogs of 6 to just over 14 years and divided them into three age groups: late adulthood (6–8 years), seniors (8–10 years), and geriatric (10 years and over).[4] There were 75 Border Collies and 110 dogs of other breeds and mixed breeds. They all took part in two experiments that were designed so the dogs did not need prior training. Owners completed a questionnaire that included the dog's participation in thirteen different types of training activities, including puppy

Senior dogs may need a little extra care, but they are still part of the family. *JEAN BALLARD*

class, obedience, agility, service dog training, hunting/nose work, trick training/dog dancing, and sheep dog training.

The first experiment tested the extent to which a social stimulus (a moving person) or non-social stimulus (a moving toy) could get and keep the dog's attention. Both senior and geriatric dogs took longer to look at the stimulus than those in late adulthood. All of the dogs looked for longer at the person than at the toy. Sustained attention declined with age and was worst in the geriatric dogs. But dogs with a high level of lifelong training kept their attention on the stimulus for longer than those with a low level of training.

The second experiment looked at selective attention. Each dog took part in a five-minute clicker training session. The experimenter called the dog to her and threw a piece of sausage on the floor. Then, every time the dog made eye contact, she clicked and threw another piece of sausage. If the dog lost interest, she crinkled the plastic bag (we all know that's a good way to get a dog's attention!). This task required the dog to switch attention from making eye contact with the person to finding the food on the floor. Unlike in humans, age did not affect selective attention in this task.

Dogs with higher scores for lifelong training and also dogs with prior experience of clicker training made eye contact faster than those with low levels of lifelong training and those with no specific experience of clicker training. Older dogs took longer to find the food on the floor, with geriatric dogs taking the longest, and this ties in with previous work on aging in dogs. There were no differences due to lifelong training in the time to find food, but dogs with prior clicker training experience were quicker to find the food than those without that experience. The clicker-trained dogs had more experience of looking for food after the click, by definition, and the researchers said they may also have had increased anticipation of food. From these results, it's not possible to separate the effects of clicker training specifically from other kinds of training, as clicker training contributed to the lifelong training scores. All of the dogs improved at the task of making eye contact during the five-minute session, which shows you can teach an old dog new tricks.

This team of researchers has also looked at the idea that computer games could be used to provide mental stimulation to older dogs. Meanwhile, studies of laboratory Beagles show extra enrichment and/or an improved diet can, at least in the short term, make a difference to cognitive aging. A clinical trial in pet dogs with canine cognitive dysfunction found nutritional supplements led to improvements in social interactions with people, fewer house-training mistakes, and less disorientation compared with dogs fed a placebo diet.[5] Another study gave dogs ages nine to eleven-and-a-half either a regular diet or a diet the scientists called "brain protection blend" that included fish oil, B vitamins, antioxidants, and L-arginine.[6] Tests showed dogs on the special diet did better at some complex learning tasks (but not on others). If you notice changes in your dog's behavior as they get older, see a vet. In the meantime, it seems that an improved diet is in the category of "can't hurt, might help" for older dogs.

Jean Donaldson of the Academy for Dog Trainers says it's important for senior dogs to have regular wellness checks with the veterinarian to ensure the dog is comfortable. "The physical comfort is really important," she said. "From a behavioral standpoint, I think I would put out a plea to not neglect the behavioral wellness of senior dogs, just because we can get away with it a little bit more. I think when dogs are young, if we do neglect to enrich their environment or give them lots of stimulation and exercise and training, we pay a price and so it prompts us to do something. Senior dogs, because they're a little bit more forgiving, sometimes I think they don't get as much as they might.

"I think the biggest mistake that I see over and over that's pretty serious," she said, "is bringing in a very young dog in some sort of attempt at companionship or because there's time available for that." She says some senior dogs love having a companion, but not all of them. "And so before making the decision to bring in another pet—dog especially, I'd say cats are less of an issue—but a dog, to be sure that we're not making the mistake of having the equivalent of a 65-year-old or 75-year-old lady having to co-house with a 15-year-old kid who's going to listen to Death Metal and have piercings and have parties in the house, and so on.

"One of these things I'm sort of sensitive about is, just because a senior dog puts up with it or just occasionally growls, that we don't say 'well, he's grumpy.' When actually we have this dog that's now senior, now maybe a little more anxious, a little more frail, that he gets some consideration, and he needs to have a veto power in whether or not another dog is brought into the family." It's a reminder that senior dogs are often easier to live with, but we still need to take their needs into account.

"THE ANIMAL SHELTER community has made tremendous strides toward reducing the number of dogs living in animal shelters; however, more work needs to be done to keep dogs with people who love them but need assistance. The world would be better for dogs if more dog owners knew about pet retention resources in their communities. Let's face it, any of us could fall on hard times, making it a struggle to care for our dogs. Fortunately, many communities have pet food pantries, low-cost veterinary services, and affordable, pet-friendly housing options for people in financial distress, but these services are not helpful unless the people who need them know about them. Some of my students in Canisius College's Anthrozoology master's program spent the past month compiling information about pet retention resources in their communities. Many found that excellent resources existed, but locating those resources typically required making multiple phone calls, sending many e-mails, and even making in-person visits to organizations. My students realized that people experiencing financial crises may not have the time and technology required to track down hard-to-find resources. This project helped my students and me realize that connecting information about pet retention resources with the dog owners who need them would be a way to make the world better for dogs."

—**CHRISTY HOFFMAN**, PhD, anthrozoologist, Canisius College

DOGS WITH SPECIAL NEEDS

SENIOR DOGS ARE not the only ones who need extra care and attention. Dogs with special needs may also need extra help, although it may not be as much as you think. Recently I met a black Labrador puppy who was just as friendly and keen to meet people as you would expect any puppy to be. It was not at all obvious he could not see, and I was told you could not tell when he was playing with other puppies either. Over the years I have met the occasional dog with vision or hearing loss and come to the conclusion that in many ways, they are just like any other dog.

Making accommodations for
hearing and vision impairment

Because very little was known about how dogs with inherited or acquired hearing or vision loss behave, a study in the *Journal of Veterinary Behavior* surveyed the owners of 461 hearing-impaired and/or vision-impaired dogs.[7] Owners completed the C-BARQ, a widely used tool to assess the behavioral traits of dogs. The results showed many similarities between dogs with a hearing or vision impairment (HVI) and those without. The non-HVI dogs were rated as more aggressive and more excitable than those with HVI. There were some differences in specific behaviors: non-HVI dogs were more likely to chase rabbits and to eat or roll in feces, whereas the HVI dogs were more likely to bark too much, lick a lot, or chew unsuitable objects. Given that barking, licking, and chewing are all ways of getting some stimulation, it seems possible these differences are due to reduced sensory input. Since the survey asked owners about any other health issues, health is not the cause of this difference.

The scientists think the HVI dogs are making up for the lack of input from their ears or eyes with behaviors that engage their other senses. This suggests owners of dogs with hearing or vision problems should make an explicit effort to make sure their dog has enough sensory input. Good ideas are enrichment with toys, including vibrating toys, Kongs, and chew toys, as well as training sessions to engage the dog's brain. Many such dogs can also attend agility, flyball, obedience, or even dog dance classes to provide enrichment and socialization opportunities. Deaf dogs can be taught to "check in" by looking at you often. The HVI dogs in the study were more likely to have had formal training, often using more hand signs and physical prompts than non-HVI dogs, and the lower levels of aggression and excitability could be for this reason. These results also suggest that dogs with HVI can make good family pets.

Making accommodations for physical impairment

Dogs that have had a limb amputated due to trauma, cancer, or some other cause are often called tripod dogs. The ones I've met have always seemed happy just going about their business being a dog. A survey of forty-four people whose dog had had a limb amputated found that forty-one of the dogs adapted very well, while three had more difficulty (of which one was said to be okay and the other two had metastases that sadly resulted in their euthanasia).[8] Almost all of the dogs had adapted to having three legs within a month of the surgery, some even faster. Half of the owners had initially not wanted to proceed with the amputation when it was advised, but in almost all of these cases the dog did much better than expected.

Dogs who need a bit of extra care, for whatever reason, can still be happy, much-loved pets. The relationship with the vet is even more important as dogs age or have other issues, so remember the advice from chapter 5 about building a relationship with a vet you like. Remember to take account of your dog's needs (and how they change) in order to keep them happy throughout their lifespan.

HOW TO APPLY THE SCIENCE AT HOME

- Keep stress to a minimum, since older dogs find it harder to cope with stress.

- Maintain a routine and continue to spend time with your dog, since older dogs still benefit from these aspects of family life.

- Feed the best-quality food you can, since older dogs have lower energy requirements but an increased need for protein.

- Take your dog to the vet; older dogs may need more frequent vet visits. Don't just put things down to old age. The vet will be able to distinguish what is normal aging and what is a condition that may need treatment.

- Ensure that dogs who are older or have special needs still get plenty of enrichment opportunities.

15

THE END
OF LIFE

. . . .

WE DID NOT know Ghost's age when we adopted him, and
as time went by his health began to decline. Although
he had ups and downs, he always perked up again, even
after the time we were told he might only have a few weeks to
live. If there's one thing all dog owners can agree on, it's that dogs
don't live long enough. There's no exact figure as to how long the
average dog lives since central data is not kept and many factors—
including dog size and breed—will influence the length of life.
Dogs often do not die due to a natural cause but by euthanasia
when it has been decided their quality of life is suffering. Euthanasia may also occur for other reasons, such as if a dog causes
serious bite injuries (when, in some cases, it may be required by
local officials).

THE LIFESPAN OF DOGS

ACCORDING TO A study in *Preventive Veterinary Medicine* that looked
at data from almost 300,000 dogs insured with the firm Anicom,
the average lifespan of a dog is 13.7 years.[1] The researchers con-
structed a life table that looked at cause of death for all the dogs
who died during the 2010 financial year (4,169 dogs). Dogs who
weigh 5 to 10 kg (11–22 pounds) have the highest average lifespan
at 14.2 years, and toy dogs who weigh less than 5 kg (11 pounds)
and dogs who weigh 10 to 20 kg (22–44 pounds) have a lifespan
of 13.8 years. The bigger the dog, the shorter the average lifespan,
with dogs of 20 to 40 kg (44–88 pounds) having a lifespan of
12.5 years, and dogs over 40 kg (88 pounds) having a lifespan of
10.6 years. The table shows the average number of years remaining for dogs of different sizes at different ages. Although for toy,
small, and medium breeds the table goes to age 17+, for large
breeds it goes to 16+, and for giant breeds to 13+.

Average remaining years of life for dogs of different breed groups at different ages

AGE OF DOG (years)	TOY BREEDS (<5 kg)	SMALL BREEDS (5–10 kg)	MEDIUM (10–20 kg)	LARGE (20–40 kg)	GIANT (>40 kg)
<1	13.8	14.2	13.6	12.5	10.6
1–2	13.0	13.4	12.8	11.7	9.8
2–3	12.0	12.4	11.9	10.7	9.0
3–4	11.1	11.5	10.9	9.8	8.1
4–5	10.1	10.5	10.0	8.9	7.2
5–6	9.2	9.6	9.1	8.0	6.5

AGE OF DOG	TOY BREEDS	SMALL BREEDS	MEDIUM	LARGE	GIANT
6–7	8.2	8.6	8.2	7.2	5.9
7–8	7.3	7.7	7.3	6.3	5.6
8–9	6.4	6.8	6.4	5.6	5.1
9–10	5.5	5.9	5.6	4.9	4.7
10–11	4.6	5.1	4.8	4.3	4.6
11–12	3.9	4.3	4.1	3.8	4.9
12–13	3.2	3.5	3.4	3.3	4.9
13–14	2.7	2.9	2.8	2.8	6.3 (age 13+)
14–15	2.1	2.2	2.2	2.7	
15–16	1.7	1.7	1.7	2.8	
16–17	1.4	1.3	1.2	3.0 (age 16+)	
17+	1.6	1.0	1.2		

Source: Data from Inoue (2015)[2]

Reprinted from Preventive Veterinary Medicine, 120(2), Mai Inoue, A. Hasegawa, Y. Hosoi, K. Sugiura, A current life table and causes of death for dogs insured in Japan, 210-218, copyright 2015, with permission from Elsevier.

The scientists pointed out that since people sometimes decide not to renew their insurance policies if they do not feel they are getting value for money (e.g., if the dog is healthy), the study likely underestimates longevity. The two most common causes of death in this study were cancer and cardiovascular disease.

In the UK, a survey of almost 14,000 owners of pedigree dogs found a life expectancy of 11 years and 3 months.[3] Only 20 percent of dogs were alive at 14 years of age, and only 10 percent at 15 years. Like the previous study, these results also show large dogs do not live as long as smaller dogs. The survey reported a few individual long-lived dogs, including a Standard Schnauzer who lived to 20 years (median age at death for this breed was almost 12), a Border Terrier who lived to 22 (median age at death was 14), and a King Charles Spaniel who lived to 23 (median age at death was 10). One of my favorite breeds of dog, the Bernese Mountain Dog, is reported as having an average lifespan of just 8 years. There are eleven breeds with a median age of 7 years or less at death, including the Great Dane, Shar Pei, Shiba Inu (Japanese), Irish Wolfhound, Bulldog, and Dogue de Bordeaux. On the happier end of things, there are fourteen breeds that live, on average, for at least 13.5 years, including the Toy Poodle, Miniature Poodle, Canaan Dog, Border Terrier, Cairn Terrier, Basenji, and Italian Greyhound.

So why are bigger dogs more likely to die young? One reason could be that larger dogs are more likely to die from specific conditions than smaller dogs are. A study published in the *Journal of Veterinary Internal Medicine* looked at the cause of death of all the dogs who had died at a veterinary teaching hospital in the US in the twenty years until 2004.[4] After excluding cases where the dog was dead on arrival or no reason was given, they had information about the deaths of almost 75,000 dogs.

Overall, the five most common causes of death were gastrointestinal, neurologic, musculoskeletal, cardiovascular, and urogenital diseases. Larger dogs are more likely to die of cancer and because of musculoskeletal or gastrointestinal issues. Smaller dogs, while less likely to get cancer, are more likely to die because of metabolic issues. Both small and large dogs die of trauma at similar rates. There are also differences due to age: infectious disease and gastrointestinal disease were more common causes of death in younger dogs, while older dogs were more likely to die from cancer or neurological disease.

A study published in the *Veterinary Journal* looked at causes of death of pet dogs in England based on data from eighty-six veterinary practices.[5] In dogs under 3 years old, behavioral issues were the main cause of death (14.7%), followed by gastro-intestinal disorders (14.5%) and road traffic accidents (12.7%). In the overall sample, the median age at death was 12 years, with tumors the main cause of death (16.5%), followed by musculo-skeletal disorders (11.3%) and neurological disorders (11.2%). After taking size into account, mixed-breed dogs lived 1.2 years longer than purebred dogs. While it's possible this is due to fewer genetic defects, it's important to remember that other factors are also likely to differ, so the cause of this difference is not known. Life expectancy varied a lot between breeds, with Miniature Poodles and Bearded Collies having the longest lifespans at 14.2 and 13.7 years respectively, and the Dogue de Bordeaux and Great Dane coming in at 5.5 and 6 years respectively.

In this study, 86.4 percent of deaths were by euthanasia. Many dogs do not die of natural causes because part of our role as guardian is to give them a timely death, without suffering. Perhaps this is the hardest part of sharing our lives with a dog.

MAKING END-OF-LIFE DECISIONS

EVEN NOW I find it hard to talk about Ghost without a frog appearing at the back of my throat. His health problems started with a tumor that had to be removed from his butt within a few weeks of us adopting him, and an intermittent general malaise. When another, different tumor was removed from his rear end four years after the first one, the vet told me that if another one came, she would not be able to operate again because of its location and the risk of fecal incontinence. When he needed to poop, he had to poop, and if we weren't there to immediately rush him outside, he could not help but have accidents in the house. This made him subdued, even though we always cleaned up the mess without a word.

Ghost's lack of energy came and went, and some days were better than others. But he still wanted his walks, even though they got slower, and he took great interest in sniffing *everything*. When we took him to one of his favorite spots down by the river, he was always reluctant to turn back towards the car, even though we were afraid if we walked too far he would be wiped out for the rest of the day. We kept his routine the same and whenever he had to have medication, we would hand-feed it hidden in cheese or a piece of salami to help it go down. He became suspicious of both cheese and salami. We would have to offer several unadulterated pieces for him to sniff cautiously and then eat, before giving him the one that contained the antibiotic or steroid. Despite everything, he seemed happy. Then something appeared on his butt again.

Sometimes people say, "You'll know when it's time," but how do you get the balance right between wanting your dog to have as long a life as possible, and not wanting them to suffer? And how do you try to consider what your dog might want when they are not able to tell you?

When a dog has a medical condition that gets gradually worse, your perception changes over time as you get used to each new normal. Euthanasia is not an easy decision. And sometimes you are left with both the knowledge of having done the right thing and the feeling of guilt that nonetheless you took a life. We can only do our best for the individual, much-loved dog who we have. This will almost always involve discussions with the vet about what is for the best, how to weigh up different treatment options (if there are any), or whether palliative care might help make life good enough for now.

A good relationship with your vet will help with these discussions. It may help to start these conversations sooner rather than later, when thinking about your dog's end of life is a less emotive issue for you, and so you are forewarned about what to consider in terms of quality of life. It may also help to plan what is right for you and your pet when the time comes; for example, should you go to the vet or would it be better if the vet came to your house (if they offer this service)?

Thinking about when is the right time for euthanasia may also make us reflect on our own mortality and that of human family members, which can be difficult to contemplate. Our own feelings are inevitably part of the decision too. On the one hand, there may be times when the dog's condition makes life difficult for the owner (such as when there are frequent messes in the house or if the dog's behavior changes as a result of their condition, perhaps making them unpredictably grumpy or difficult). Some people may wish to euthanize an animal sooner, while others may prefer to wait, even if the animal no longer has a good quality of life.[6] Discussions about what to do are based on our own ethics and our assessment—insofar as we can tell—of the dog's quality of life.

Gather evidence

Quality-of-life scales can be used to help make a decision about ways in which you might improve your dog's quality of life and whether your dog has a life worth living. A review in the *Veterinary Journal* notes that many of these scales are designed for specific health conditions such as osteoarthritis, spinal cord injuries, or pain related to cancer.[7] As well, few of the questionnaires actually define quality of life. It can be specific to the effects of a medical condition, but there is also a broader concept of general quality of life. While more research is clearly needed, the review identifies a number of questionnaires that dog owners or veterinarians might find useful in evaluating quality of life for a particular dog.

Palliative care, also known as hospice care, involves keeping a dog as comfortable as possible in their home. Some veterinary practices specialize in palliative care and will make house calls to discuss ongoing quality of life and what can be done to improve it, as well as any modifications to the home that might help (such as a ramp to replace steps for dogs with mobility issues). Palliative care may allow a family to keep their dog comfortable for a few more days or weeks at home while they can say goodbye to their pet.[8] One study of dogs undergoing palliative radiation therapy for tumors found 79 percent of owners were happy with their decision for the dog to be treated, and 78 percent reported that it improved their pet's quality of life; however, 18 percent believed the treatment might be curative.[9] There is a moral balance between prolonging life and potentially prolonging suffering. Ultimately, palliative care will often still end in euthanasia when that is considered the best choice for the dog.

Veterinarian Dr. Adrian Walton says people should think about two components of euthanasia if they have an elderly or chronically ill pet: "There is the quality of life for the pet and there is also the quality of life for the owner. One of the things that I say as a veterinarian is my job is not to save the life of your pet; my job is to save your relationship with your pet." He explained, "There are going to be times, especially when we're talking about end-of-life situations, so for instance kidney disease, diabetes, heart disease, where the medications wind up causing animals to say, pee all over the place, or become disorientated, or they no longer seem to be the pet. And what I do find is that sometimes owners get so worked up and so stressed over this stage in a pet's life that it negatively impacts them. And beyond that it also impacts the pets . . . So one of the things we find with kidney failure dogs is that dogs are often isolated in a laundry room or bathroom or garage, so that people can contain the peeing, as it were. But then all of a sudden the pet is disconnected from the family and that's when we see a lot of cases of unintentional neglect. So one of the things I'd be saying is if you're finding that this is negatively impacting your life, then euthanasia's on the table. But as for the pet, the simplest way is, when the bad days outnumber the good, then certainly euthanasia's on the table."

Know that decision-making can be difficult

Sometimes the decision whether to try (or continue) treatment or to euthanize can be a tricky one. In some cases it will be obvious, as with a sudden catastrophic decline or a road traffic accident, but in other cases there may be more than one right answer from an ethical point of view. A study published in *Acta Veterinaria Scandinavica* looked at precisely this kind of situation.[10] Scientists interviewed twelve owners of pets where the notes in the vet's file indicated that euthanasia had been discussed; in some cases, the animal had been euthanized in the previous eight months, and in other cases, the animal was still alive and receiving treatment. The dogs had conditions like epilepsy, diabetes, asthma, dementia, and cancer.

People found the decision easier if the pet was suffering or deteriorated suddenly, if the person was struggling, or if there were no more options for treatment. It was harder when there was a gradual deterioration or when better days alternated with worse days, making it hard to know when to draw a line, or why to make the decision on any particular day compared with the previous day or any number of other similar days.

Another thing people said made the decision difficult was lack of knowledge, whether it was understanding the condition, finding it hard to take in what the vet said about it, or finding it hard to evaluate the dog's welfare. People reported conflicts when it was hard to know whether to try treatment if it might have side effects or potentially might only make the dog suffer for longer. People also said they found it hard to know whether it was right to consider their own needs and struggles with the dog's condition, or only to consider the dog's perspective. The responsibility of the decision itself also weighed heavily.

Obviously the vet's expert knowledge was needed in helping to make the decision, but also people found that it helped if the vet guided them, as it helped to share the responsibility for the decision. Thus, the vet could potentially really help, not only in terms of helping to make the right decision for the dog's welfare but also in helping the dog's guardian cope with the difficulty of making the decision.

This view is supported by research in *Anthrozoös* on people's grief after the loss of a pet.[11] Feelings of sorrow, grief, anger, and guilt are all common feelings after a pet has been euthanized. Feelings of guilt seem to be less if people can say the decision they made was the one that was right for their pet at the time, and if veterinarians and their staff have supported them in making that decision. Feelings of anger and guilt were less when the pet had had cancer, and feelings of anger were greater in situations where the animal had died a sudden death. When the pet's death is sudden, or when the owner has a very strong attachment to the pet, the person may need additional support after the pet's death. These days, grief counseling is increasingly available for those who need it, including from specialist pet loss counselors.

Social support from friends and family is also important. People who have recently lost a pet (whether by euthanasia or natural death) show increased levels of stress and lower physical, psychological, and relationship aspects of quality of life. A study in PLOS

ONE looked at the stress and quality of life of owners euthanizing their dogs and found that people often experience a lack of social support.[12] When someone loses a person, people assume social support is needed, but they do not always realize the depth of feelings on loss of a pet.

While some people feel the need to get another pet right away, others need to grieve for a long time before they are ready. There are no right answers, and it's up to you what you think is best for you and your family.

EUTHANASIA: WHAT TO EXPECT

ONE THING MANY people worry about is whether or not to be there at the time of their dog's death. "That is completely up to the owner," said Dr. Walton. "Some people need to be present; some people don't. The truth is the animal doesn't care one way or the other because the animal doesn't know what's going on. One minute they're there; one minute they're not . . . So I tell people that is something that is an individual decision for you."

For many people, euthanizing a pet is their first experience of death, so I asked Walton what people need to know about the process itself and what they should expect. "People are not aware of how fast it is," he said. "Basically I think what people are expecting is that the dog slowly disappears over a period of 5 minutes, when actually it's more like 30 seconds. The other thing is . . . people are not aware that after death the eyes don't shut. Most people actually think from watching television that the eyes close when the animal dies, when in fact they stay open and basically are staring off into space."

He added, "There are certain reactions that the body goes through that people are not aware of. The first one is called agonal breathing, and what that is, is when the heart stops because the brain is no longer sending it a signal, carbon dioxide builds up in the bloodstream and there's an actual reflex, with little receptors in the carotid artery, that basically will force the chest to expand. And so what you get is this." He took two sharp, noisy intakes of breath. "And that can go on for several minutes and people aren't aware of that.

"And then, of course, there's also the fact that they may stretch, you may see muscles twitch, you may see loss of control of the bowel or bladder. A lot of times I'll do a home euthanasia and the dog is lying on a person's Persian carpet and you're trying to explain to them that we need to move the dog onto a blanket or onto some plastic, to understand that."

Finally, he noted, for people whose dogs have died overnight, there is rigor mortis. "A lot of people do not know that your animal will become stiff."

PREPARING PETS FOR THE DEATH OF A COMPANION

IT WOULD SEEM only natural for dogs to grieve when one of their animal companions passes on. A survey of pet owners asked people to report on their dog or cat's behavior when another dog or cat in the household passed away.[13] Sixty percent of the dogs were reported to keep checking the spot where their housemate would normally be resting, and 61 percent were said to want more affection or to be clingy or needy. Thirty-five percent of dogs ate less and 31 percent ate more slowly following their companion's death, and a similar percentage were said to spend more time sleeping. All of this is consistent with the dog experiencing grief, and interestingly, the same changes were seen whether it was a dog or cat who had passed away. In this study, 58 percent of the dogs had seen the body after their companion died, and 73 percent of those had taken time to sniff it. But there were no differences in the reported behavioral changes whether or not they had seen the dead body, showing this is not necessary for them to experience grief.

Ghost's ashes live in a pot near the television, and his paw print is on a shelf where I see it many times a day. I think of him often. You know how people sometimes speak of having a "heart dog," your canine soul mate? To me, Ghost was just the most incredible creature we could have brought to live in our home. I miss his happy woo-woos, the way he would sit and gaze at me when I was watching TV, all the walks he accompanied me on. I miss seeing his tail curl up and over his back, and the way he was so alert to wildlife. I miss the way he would come and put his big head on my knee when he wanted me to pet him. I miss the feel of my fingers in his thick fur. Most of all, I miss his companionship.

Through those early days without him, Bodger really helped. He was sad himself, but he still needed to be walked and fed and petted and to play tug. He has grown used to being the only dog in the house. Right now, he is looking out of the window, but he has half an eye on me, because he knows it's almost time for his walk. I am looking at the rain and wondering if he will want to walk or if he will choose to stay home. Every dog is an individual, and in life or in death, we try to honor them as best we can.

PLANNING FOR THE UNEXPECTED

OF COURSE, IT'S also important to plan for if we are hospitalized or predecease our pets. This may include letting friends and family know what you want as well as speaking to your lawyer to make decisions about wills, trusts, and powers of attorney. In

Texas, the Stevenson Companion Animal Life-Care Center cares for pets in a home-like environment after their owners pass. Some organizations provide foster care for pets when their owners are hospitalized or unable to care for them for other reasons (such as fleeing domestic violence, as with the Dogs Trust Freedom Project). When you designate someone to look after your pet, keep in touch with them as their circumstances or your pet's needs may change over time. Try to pick someone who has experience with caring for pets and who will respect your wishes. As well, a trusted friend or neighbor should have a set of house keys and know how to care for your dog in case you are taken to hospital or delayed getting home because of an emergency. A sticker on your door or window can tell first responders how many pets live in your home.

"ONE THING THAT would make the world better for dogs is if we could address the underlying factors that contribute to their relinquishment. Behavioral concerns and housing issues are among the main reasons why people give up their pets. I wish people would acquire dogs not based on physical appearance or on the latest trends, but rather, by going through a reputable source to find the right dog that matches one's personality and lifestyle. Reputable sources can provide support for dog owners, with many going as far as partnering with local dog trainers and veterinarians. Finally, to keep dogs and their owners together, I wish rental-housing policy would consider pets as part of people's families. Dogs are often banned from rental housing, which impacts people's capacities to keep and care for them."

—TARYN GRAHAM, PhD, research associate, York University, and founder of PAWsitive Leadership

It is also a good idea to think about emergency planning for your pet. This includes thinking about the kinds of emergencies you might have to face (such as house fires, wildfires, earthquakes, and hurricanes) and where there are pet-friendly hotels and shelters in your neighborhood. An emergency kit for dogs should include a copy of vaccination records, any required medication, harness and leash, dog toys, food and water (enough for 3–7 days) as well as bowls, poop bags, dish soap and cleaning supplies, and blankets. Crate training your dog is a good idea in case they have to be in a crate in a shelter (e.g., if you live somewhere prone to wildfires or hurricanes). If you have to evacuate, take your dog with you. Training is helpful; after the magnitude 9.0 earthquake in Japan in 2011, amongst people who were able to evacuate with their pet, 46 percent had trained and socialized them; in comparison, amongst those who had to leave their pet behind, only 26 percent had trained and socialized them.[14] Identification for your pet (microchips, tattoos, and your phone number on their collar) will help you be reunited if separated. You should also think about safe restraint for your dog in vehicles, such as a crash-tested harness, crate, or barrier (in line with legal requirements where you live) to protect your dog in the event of a motor vehicle accident. Advance planning is essential, but disaster preparedness is not just about emergency supplies, it's also about giving your dog the skills to cope with normal, everyday life.

HOW TO APPLY THE SCIENCE AT HOME

- If your dog has a chronic condition, see if there are adjustments you can make to maintain good quality of life. Discuss euthanasia with your vet well ahead of time so that you know what signs to look out for that might indicate a reduced quality of life, and give some thought to how to manage euthanasia (e.g., at home or at the clinic).

- Think in advance about how you would make an end-of-life decision. Quality-of-life scales can be helpful in making the decision, but often apply only to specific conditions. A more general rule, as suggested by Dr. Walton, is when the bad days outnumber the good.

- Don't be afraid to ask your vet for advice or to explain the pros and cons of different treatment options (if available) to help you make the decision.

- Be sensitive to companion animals. Any dogs remaining in the home may grieve for their lost companion and want more affection and attention. If a pet of another species passes away, such as a cat, your dog may grieve for them too.

- Make a plan for your pet(s) in the event you become ill or pre-decease them.

- Make an emergency plan for your family (including pets). Find out which hotels near you are pet-friendly, and have a grab-bag of emergency supplies for your dog in case you need to leave home in an emergency.

16

SAFE DOGS, HAPPY DOGS

·················

'VE BEEN THINKING about what Bodger's ideal life would be like. My husband and I wouldn't go out except to take him for a nice walk somewhere fairly quiet but with enough other people and dogs to provide some interest. The sun would be shining and there would be no more than a little breeze. All the people he met would want to pet him and expect to be licked on the lips. The rest of the time, we would stay home with him to keep him company.

He does like his kibble but I'm sure he would prefer home-cooked meals, perhaps steak for breakfast and something involving tripe, cheese sauce, and eggs for his dinner. There would be lots of little pieces of cheese and sausage during the day to keep his energy levels and motivation up. He would meet up with his doggy friends for an afternoon walk around the neighborhood and find something smelly like a dead mouse to roll on. There would be some bear poop, and the bear would have been eating fruit—apples, say, or plums—so he could have a good sniff and then take a bite of this delicacy. After the walk we would come home for cuddles on the settee.

Of course there would be many games of tug and perhaps a few minutes of fetch too (but nothing too exhausting). He would like a little grooming but not too much at once. The cats would wander round the house a lot so he could follow them and try to herd them back together or into another room. There would be no owls in the world. And at night he would sleep on the bed with us and the cats, although the cats would stay at a respectful distance and not get up to any mischief.

Ghost's ideal life would have included more walks and opportunities to roam far through the forest and forage for salmonberries, mice, and perhaps a squirrel before returning home to the same meal of tripe, cheese sauce, and eggs. The cats would have been safely shut away so as not to provide any temptation, since it's hard to have a friend who is also food. There would have been a nice long grooming session—at least half an hour—to help him wind down at the end of the day. And we would never have gone out and left him at home alone.

Unfortunately, this perfect life is not achievable. Sometimes it's necessary to go out, although Bodger is used to this and copes well with it now. I ask him to leave the bear poop, if I spot it in time, and I drop treats on the ground as a thank you—but I see the regretful look he gives the poop as he walks away, and the way he remembers it and wants to return to the same spot on his next few walks. And sometimes, frankly, the cats are too exciting, especially when they jump to the top of the cat tree and don't come down or when Melina—still like a kitten—climbs the curtains in the bedroom. I can't really bear to cook tripe, so he has to make do with the canned variety, although apparently this is pretty good too. And of course, in Bodger's perfect world—just as in mine—Ghost would still be with us.

This book has outlined what it means for dogs to be happy and how canine science contributes to our understanding of what pet dogs need. Although it is always said that more research is needed, it is, of course, true. In particular, it would be nice to see more studies with a larger number of participating dogs, longitudinal studies to track differences over time, and a good balance between experimental and naturalistic studies.

When thinking about animal welfare, Prof. David Mellor told me that we need to both minimize negative experiences *and* provide opportunities for positive experiences. For example, if an animal is in pain, they will not want to play or engage with the people and animals in their household, so we need to make sure that the pain is not intolerable and also realize it may interfere with the dog's ability to enjoy positive experiences. This means positive and negative experiences are both important when considering the dog's welfare.

An illustration of some of the main factors that affect pet dog welfare, arranged in terms of the Five Domains, is shown in the table.

Good welfare for happy dogs

NUTRITION	ENVIRONMENT	HEALTH	BEHAVIOR
• Good food • Water	• Safely enclosed • Nice dog beds • A safe place to relax away from children • Appropriate bylaws and legal framework	• Good health • Low-stress veterinary care • Training for body handling and grooming • Breeding for good health and to avoid hereditary conditions • Walks and other exercise	• Positive experiences during the sensitive period for socialization • Companionship (with or apart from other dogs, humans) • Opportunities to play • Reward-based training • Opportunities for enrichment

MENTAL HEALTH
• A sense of safety • Prompt help for behavior issues • Choices

HOW TO PROMOTE GOOD ANIMAL WELFARE

THERE ARE A number of things we can do to ensure our dogs are happy:

- Help your dog feel safe (chapters 3, 7, and 13)
- Learn to recognize signs of fear, anxiety, and stress (chapters 1 and 13)
- Socialize puppies during the sensitive period for socialization (chapter 2)
- Give your dog choices (chapter 2)
- Train using reward-based methods (chapters 3 and 4)
- Ensure your dog gets regular walks and other exercise (chapter 9)
- Find a vet who uses low-stress handling techniques (chapter 5)
- Ask your vet about your dog's weight (chapter 11)
- Provide opportunities for enrichment, from "sniffaris" to play (chapters 6 and 10)
- Make changes to reflect your dog's needs as they age (chapter 14)

And remember, every dog is an individual, and it's up to you to find what works for your dog.

The first part of making dogs happy is learning more about them. These days, it's increasingly easy to find information about dogs and how to care for them. The trouble is that much of this information is erroneous and things are changing all the time as research into all aspects of dogs' behavior, cognition, and welfare continues apace. Given the poor quality of so much information, one thing we can all do to help dogs is share good books, blogs, and websites when we find them. Follow my own blog, *Companion Animal Psychology*, to keep up with the latest developments and for suggestions of other blogs (and books) to read too. And consider having your dog take part in canine science if you wish, either online from a distance via groups like Darwin's Ark, Dognition, and Generation Pup, or quite possibly in person at a university in a town near you. These are exciting times for dogs and our relationship with them. As this book shows, there are many things dog owners are getting right, but also lots of room for improvement.

The second part of making dogs happier is putting what we've learned about canine behavior and animal welfare and what you know about your own dog's needs and preferences into action. At the end of the book, I have included a checklist in case you want to apply some of the ideas to your own dog(s). In making the list, I have drawn on the science detailed in this book as well as my own experience from training dogs. It includes questions about what makes your dog happy—and what might make them afraid or stressed. It is not intended as a substitute for advice from

Dogs' needs change through their lifespan, but they can still be happy at any age.
JEAN BALLARD

a professional, so if you have any concerns, see your vet or a qual-ified dog trainer or behaviorist.

Bodger is an almost-constant presence in my life and we spend most of our waking hours together. At night, he sleeps on a dog bed at the foot of the bed. Sometimes when I write, he comes and lounges in my study, sometimes falling asleep so close that I can no longer move my chair. He picks the spot that used to be Ghost's, so it's one way in which he has adjusted to Ghost's absence. But I've noticed another way in which he hasn't: at meal times, he will trot the long way round to his food bowl, as if Ghost were here and eating in the space where he always ate. I'm sure it's just because it became a habit to let Ghost eat in peace (whilst secretly watching for leftovers). But it shows that, even when gone, there is still a trace of the dog we loved. The impact dogs have on our lives is tremendous. We owe it to them to do what we can to make them happy.

CHECKLIST FOR A
HAPPY DOG

T HIS CHECKLIST IS designed to help you think about some of
the ideas in this book in relation to your own dog. It is not
a scientifically validated instrument, nor is it a substitute for
a professional opinion. If you have any concerns about your dog,
consult a veterinarian, dog trainer, or animal behaviorist.

Answer each question yes or no. The more "yes" answers the
better. For the "no" answers, troubleshoot the situation to see if
you should make changes.

Dog's Name:
Age:
Breed:

		YES/ NO	CHAPTER
1	My dog has a safe space to go if they feel stressed (e.g., dog bed, crate).		1, 13
2	Everyone in the household respects the dog's safe space and leaves them alone when they are there.		1, 13
3	My dog has a daily routine.		1, 13
4	Everyone in the household is consistent about the rules that apply to the dog.		1, 13
5	My dog has a choice of whether or not to interact with each member of the family.		1, 13
6	Children are never allowed to approach my dog when the dog is sitting or lying down.		8
7	If very young children want to pet my dog, an adult guides their hand to ensure they are gentle.		8
8	My dog has at least one walk a day.		9
9	My dog gets sniffing opportunities on the walks.		9
10	My dog is not regularly left alone for more than 4 hours at a time (or if they are, I make arrangements for a dog walker, family member, or neighbor to visit).		7

		YES/ NO	CHAPTER
11	My dog has annual vet visits (or as determined by the vet).		5
12	My dog is up to date on vaccinations as recommended by their vet.		5
13	My dog has regular treatments for parasites as recommended by their vet.		5
14	My dog is a healthy weight (if in doubt, ask your vet).		5, 11
15	If my dog is friendly to other dogs, they get opportunities to socialize with other similar dogs.		6
16	If my dog finds socializing with other dogs stressful, I do not make them do it.		6
17	My dog has opportunities to play with food toys.		10
18	My dog has chew toys.		10
19	My dog has opportunities to use their nose (e.g., snuffle mat, sniffaris, etc.).		10
20	No one in the household uses positive punishment (e.g., yelling, hitting, shock collar, prong collar) to train the dog.		3, 4, 13
21	The dog is regularly taught new things using positive reinforcement.		3, 4, 13
22	When traveling by car, the dog is safely restrained in line with legal requirements where I live.		15

		YES/NO	CHAPTER
23	My dog is friendly to other people.		3, 4, 13
24	My dog is friendly to family members.		3, 4, 13
25	I look out for signs of stress in my dog (e.g., lip licking, looking away, trembling, hiding, seeking out people) and intervene to help if needed.		1
26	I have made plans for my dog in case something happens to me.		15
27	My dog is included in my family's emergency planning.		15
28	My dog has bed(s) they like to sleep in.		12
29	If needed, I have adapted the environment to accommodate my dog's special needs.		14
	My dog's favorite toys are:		
	My dog's favorite games are:		
	My dog's favorite places to walk are:		

NOTES

Chapter 1 Happy Dogs

1. American Pet Products Association, "Pet Industry Market Size and Ownership Statistics," Accessed June 17, 2019, americanpetproducts.org/press_industrytrends.asp.

2. Statista, "Number of dogs in the United States from 2000 to 2017 (in millions)," Accessed August 8, 2018, statista.com/statistics/198100/dogs-in-the-united-states-since-2000/; Canadian Animal Health Institute, "Latest Canadian pet population figures released," January 28, 2019, cahi-icsa.ca/press-releases/latest-canadian-pet-population-figures-released; Pet Food Manufacturers' Association, "Pet Population 2018," Accessed August 8, 2018, pfma.org.uk/pet-population-2018.

3. M. Wan, N. Bolger, and F.A. Champagne, "Human perception of fear in dogs varies according to experience with dogs," *PLoS ONE* 7, no. 12 (2012): e51775.

4. Chiara Mariti et al., "Perception of dogs' stress by their owners," *Journal of Veterinary Behavior* 7, no. 4 (2012): 213–219; Emily J. Blackwell, John W.S. Bradshaw, and Rachel A. Casey, "Fear responses to noises in domestic

dogs: Prevalence, risk factors and co-occurrence with other fear related behaviour," *Applied Animal Behaviour Science* 145, no. 1–2 (2013): 15–25.

5. S.D.A. Leaver and T.E. Reimchen, "Behavioural responses of *Canis familiaris* to different tail lengths of a remotely-controlled life-size dog replica," *Behaviour* (2008): 377–390.

6. Marc Bekoff, *The Emotional Lives of Animals: A Leading Scientist Explores Animal Joy, Sorrow, and Empathy—and Why They Matter* (Novato, CA: New World Library, 2010); Jonathan Balcombe, *What a Fish Knows: The Inner Lives of Our Underwater Cousins* (New York: Scientific American/Farrar, Straus and Giroux, 2016).

7. Jaak Panksepp, "Affective consciousness: Core emotional feelings in animals and humans," *Consciousness and Cognition* 14, no. 1 (2005): 30–80.

8. National Archives, "Farm Animal Welfare Council Five Freedoms," 2012, webarchive.nationalarchives.gov.uk/20121010012427/http://www.fawc.org.uk/freedoms.htm.

9. John Webster, "Animal welfare: Freedoms, dominions and 'a life worth living,'" *Animals* 6, no. 6 (2016): 35.

10. People's Dispensary for Sick Animals, "Animal Wellbeing PAW Report," 2017, pdsa.org.uk/media/3291/pdsa-paw-report-2017_printable-1.pdf.

11. David J. Mellor, "Updating animal welfare thinking: Moving beyond the 'five freedoms' towards 'a life worth living,'" *Animals* 6, no. 3 (2016): 21; David J. Mellor, "Moving beyond the 'five freedoms' by updating the 'five provisions' and introducing aligned 'animal welfare aims,'" *Animals* 6, no. 10 (2016): 59.

12 David J. Mellor, "Operational details of the Five Domains Model and its key applications to the assessment and management of animal welfare," *Animals* 7, no. 8 (2017): 60.

13. Alexander Weiss, Mark J. Adams, and James E. King, "Happy orang-utans live longer lives," *Biology Letters* (2011): rsbl20110543.

14. Lauren M. Robinson et al., "Happiness is positive welfare in brown capu-chins (*Sapajus apella*)," *Applied Animal Behaviour Science* 181 (2016): 145–151; Lauren M. Robinson et al., "Chimpanzees with positive welfare are happier, extraverted, and emotionally stable," *Applied Animal Behaviour Science* 191 (2017): 90–97.

15. Nancy A. Dreschel, "The effects of fear and anxiety on health and life-span in pet dogs," *Applied Animal Behaviour Science* 125, no. 3–4 (2010): 157–162.

16. American Society for the Prevention of Cruelty to Animals (ASPCA), "Facts about US animal shelters," Accessed April 7, 2018, aspca.org/animal-homelessness/shelter-intake-and-surrender/pet-statistics.

17. American Veterinary Society of Animal Behavior (AVSAB), "Position statement on puppy socialization," 2008, avsab.org/wp-content/uploads/2018/03/Puppy_Socialization_Position_Statement_Download_-_10-3-14.pdf.

18. Dan G. O'Neill et al., "Longevity and mortality of owned dogs in England," *The Veterinary Journal* 198, no.3 (2013): 638–643.

19. American Humane Association, "Keeping pets (dogs and cats) in homes: A three-phase retention study. Phase II: Descriptive study of post-adoption retention in six shelters in three US cities," 2013, americanhumane.org/publication/keeping-pets-dogs-and-cats-in-homes-phase-ii-descriptive-study-of-post-adoption-retention-in-six-shelters-in-three-u-s-cities/; BBC, "RSPCA launches Puppy Smart campaign," February 1, 2011, news.bbc.co.uk/local/cornwall/hi/people_and_places/nature/newsid_9383000/9383583.stm.

20. People's Dispensary for Sick Animals, "Paw Report 2018," pdsa.org.uk/media/4371/paw-2018-full-web-ready.pdf; Kate M. Mornement et al., "Evaluation of the predictive validity of the Behavioural Assessment for Re-homing K9's (B.A.R.K.) protocol and owner satisfaction with adopted dogs," *Applied Animal Behaviour Science* 167 (2015): 35–42.

Chapter 2 Getting a Dog

1. Lee Alan Dugatkin and Lyudmila Trut, *How to Tame a Fox (and Build a Dog): Visionary Scientists and a Siberian Tale of Jump-Started Evolution* (Chicago: University of Chicago Press, 2017).

2. Bridget M. Waller et al., "Paedomorphic facial expressions give dogs a selective advantage," *PLoS ONE* 8, no. 12 (2013): e82686.

3. Stefano Ghirlanda, Alberto Acerbi, and Harold Herzog, "Dog movie stars and dog breed popularity: A case study in media influence on choice," *PLoS ONE* 9, no. 9 (2014): e106565.

4. Stefano Ghirlanda et al., "Fashion vs. function in cultural evolution: The case of dog breed popularity," *PLoS ONE* 8, no. 9 (2013): e74770.

5. Harold A. Herzog, "Biology, culture, and the origins of pet-keeping," *Animal Behavior and Cognition* 1, no. 3 (2014): 296–308.

6. Harold A. Herzog and Steven M. Elias, "Effects of winning the Westminster Kennel Club Dog Show on breed popularity," *Journal of the American Veterinary Medical Association* 225, no. 3 (2004): 365–367.

7. Kendy T. Teng et al., "Trends in popularity of some morphological traits of purebred dogs in Australia," *Canine Genetics and Epidemiology* 3, no. 1 (2016): 2; Terry Emmerson, "Brachycephalic obstructive airway syndrome: a growing problem," *Journal of Small Animal Practice* 55, no. 11 (2014): 543–544.

8. American Kennel Club, "Most popular dog breeds of 2018 (2019),"
March 20, 2019, akc.org/expert-advice/news/most-popular-dog-
breeds-of-2018/; Canadian Kennel Club, "Announcing Canada's top
10 most popular dog breeds of 2018," January 18, 2019, ckc.ca/en/
News/2019/January/Announcing-Canada-s-Top-10-Most-Popular-
Dog-Breeds; Kennel Club, "Top twenty breeds in registration
order for the years 2017 and 2018," 2019, thekennelclub.org.uk/
media/1160202/2017-2018-top-20.pdf.

9. Peter Sandøe et al., "Why do people buy dogs with potential welfare
problems related to extreme conformation and inherited disease? A representative study of Danish owners of four small dog breeds," *PLoS ONE*
12, no. 2 (2017): e0172091.

10. R.M.A. Packer, A. Hendricks, and C.C. Burn, "Do dog owners perceive
the clinical signs related to conformational inherited disorders as 'normal' for the breed? A potential constraint to improving canine welfare,"
Animal Welfare-The UFAW Journal 21, no. 1 (2012): 81.

11. R.M.A. Packer, D. Murphy, and M.J. Farnworth, "Purchasing popular
purebreds: investigating the influence of breed-type on the pre-
purchase motivations and behaviour of dog owners," *Animal Welfare-
The UFAW Journal* 26, no. 2 (2017): 191–201.

12. M. Morrow et al., "Breed-dependent differences in the onset of fear-
related avoidance behavior in puppies," *Journal of Veterinary Behavior* 10,
no. 4 (2015): 286–294.

13. D. Freedman, J. King, and O. Elliot, "Critical period in the social
development of dogs," *Science* 133, no. 3457 (1961): 1016-1017;
C. Pfaffenberger and J. Scott, "The relationship between delayed
socialization and trainability in guide dogs," *The Journal of Genetic Psychology* 95, no. 1 (1959): 145–155; J. Scott and M. Marston, "Critical

periods affecting the development of normal and mal-adjustive social behavior of puppies," *The Pedagogical Seminary and Journal of Genetic Psychology* 77, no. 1 (1950): 25–60.

14. James Serpell, Deborah L. Duffy, and J. Andrew Jagoe, "Becoming a dog: Early experience and the development of behavior" in *The Domestic Dog: Its Evolution, Behavior and Interactions with People*, ed. James Serpell (Cambridge: Cambridge University Press, 2017); John Bradshaw, *In Defence of Dogs: Why Dogs Need Our Understanding* (London: Allen Lane, 2011).

15. F. McMillan et al., "Differences in behavioral characteristics between dogs obtained as puppies from pet stores and those obtained from noncommercial breeders," *Journal of the American Veterinary Medical Association* 242, no. 10 (2013): 1359–1363.

16. Federica Pirrone et al., "Owner-reported aggressive behavior towards familiar people may be a more prominent occurrence in pet shop-traded dogs," *Journal of Veterinary Behavior* 11 (2016): 13–17.

17. Franklin D. McMillan, "Behavioral and psychological outcomes for dogs sold as puppies through pet stores and/or born in commercial breeding establishments: Current knowledge and putative causes," *Journal of Veterinary Behavior* 19 (2017): 14–26.

18. C. Westgarth, K. Reevell, and R. Barclay, "Association between prospective owner viewing of the parents of a puppy and later referral for behavioural problems," *Veterinary Record* 170, no. 20 (2012): 517.

19. Helen Vaterlaws-Whiteside and Amandine Hartmann, "Improving puppy behavior using a new standardized socialization program," *Applied Animal Behaviour Science* 197 (2017): 55–61.

20. Kate M. Mornement et al., "Evaluation of the predictive validity of the Behavioural Assessment for Re-homing K9's (B.A.R.K.) protocol and

owner satisfaction with adopted dogs," *Applied Animal Behaviour Science* 167 (2015): 35–42.

21. Sophie Scott et al., "Follow-up surveys of people who have adopted dogs and cats from an Australian shelter," *Applied Animal Behaviour Science* 201 (2018): 40–45.

Chapter 3 How Dogs Learn

1. Pamela Joanne Reid, *Excel-erated Learning: Explaining in Plain English How Dogs Learn and How Best to Teach Them* (Berkeley, CA: James & Kenneth Publishers, 1996).

2. Enikő Kubinyi, Péter Pongrácz, and Ádám Miklósi, "Dog as a model for studying conspecific and heterospecific social learning," *Journal of Veterinary Behavior* 4, no. 1 (2009): 31–41.

3. J.M. Slabbert and O. Anne E. Rasa, "Observational learning of an acquired maternal behaviour pattern by working dog pups: An alternative training method?" *Applied Animal Behaviour Science* 53, no. 4 (1997): 309–316.

4. Claudia Fugazza and Ádám Miklósi, "Should old dog trainers learn new tricks? The efficiency of the Do as I Do method and shaping/clicker training method to train dogs," *Applied Animal Behaviour Science* 153 (2014): 53–61.

5. Dorit Mersmann et al., "Simple mechanisms can explain social learning in domestic dogs (*Canis familiaris*)," *Ethology* 117, no. 8 (2011): 675–690.

6. Zazie Todd, "Barriers to the adoption of humane dog training methods," *Journal of Veterinary Behavior* 25 (2018): 28–34.

7. Emily J. Blackwell et al., "The relationship between training methods and the occurrence of behavior problems, as reported by owners, in

a population of domestic dogs," *Journal of Veterinary Behavior* 3, no. 5 (2008): 207–217.

8. Blackwell, "Relationship between training methods."

9. Christine Arhant et al., "Behaviour of smaller and larger dogs: Effects of training methods, inconsistency of owner behaviour and level of engagement in activities with the dog," *Applied Animal Behaviour Science* 123, no. 3–4 (2010): 131–142.

10. Nicola Jane Rooney and Sarah Cowan, "Training methods and owner-dog interactions: Links with dog behaviour and learning ability," *Applied Animal Behaviour Science* 132, no. 3–4 (2011): 169–177.

11. Stéphanie Deldalle and Florence Gaunet, "Effects of 2 training methods on stress-related behaviors of the dog (*Canis familiaris*) and on the dog-owner relationship," *Journal of Veterinary Behavior* 9, no. 2 (2014): 58–65.

12. Meghan E. Herron, Frances S. Shofer, and Ilana R. Reisner, "Survey of the use and outcome of confrontational and non-confrontational training methods in client-owned dogs showing undesired behaviors," *Applied Animal Behaviour Science* 117, no. 1–2 (2009): 47–54.

13. G. Ziv, "The effects of using aversive training methods in dogs—a review," *Journal of Veterinary Behavior* 19 (2017): 50–60.

14. American Veterinary Society of Animal Behavior (AVSAB), "The AVSAB position statement on the use of punishment for behavior modification in animals," 2007, avsab.org/wp-content/uploads/2018/03/Punishment_Position_Statement-download_-_10-6-14.pdf.

15. Jonathan J. Cooper et al., "The welfare consequences and efficacy of training pet dogs with remote electronic training collars in comparison to reward based training," *PLoS ONE* 9, no. 9 (2014): e102722.

16. Nicole S. Starinsky, Linda K. Lord, and Meghan E. Herron, "Escape rates and biting histories of dogs confined to their owner's property through the use of various containment methods," *Journal of the American Veterinary Medical Association* 250, no. 3 (2017): 297–302.

17. Sylvia Masson et al., "Electronic training devices: Discussion on the pros and cons of their use in dogs as a basis for the position statement of the European Society of Veterinary Clinical Ethology," *Journal of Veterinary Behavior* 25 (2018): 71–75.

18. Carlo Siracusa, Lena Provoost, and Ilana R. Reisner, "Dog- and owner-related risk factors for consideration of euthanasia or rehoming before a referral behavioral consultation and for euthanizing or rehoming the dog after the consultation," *Journal of Veterinary Behavior* 22 (2017): 46–56.

19. Juliane Kaminski, Josep Call, and Julia Fischer, "Word learning in a domestic dog: Evidence for fast mapping," *Science* 304, no. 5677 (2004): 1682–1683; John W. Pilley, and Alliston K. Reid, "Border collie comprehends object names as verbal referents," *Behavioural Processes* 86, no. 2 (2011): 184–195.

20. Rachel A. Casey et al., "Human directed aggression in domestic dogs (*Canis familiaris*): Occurrence in different contexts and risk factors," *Applied Animal Behaviour Science* 152 (2014): 52–63.

21. Ai Kutsumi et al., "Importance of puppy training for future behavior of the dog," *Journal of Veterinary Medical Science* 75, no. 2 (2013): 141–149.

22. American Veterinary Society of Animal Behavior (AVSAB), "AVSAB position statement on puppy socialization," 2008, avsab.org/wp-content/uploads/2019/01/Puppy-Socialization-Position-Statement-FINAL.pdf

23. J.H. Cutler, J.B. Coe, and L. Niel, "Puppy socialization practices of a sample of dog owners from across Canada and the United States," *Journal of the American Veterinary Medical Association* 251, no. 12 (2017): 1415–1423.

Chapter 4 Motivation and Technique

1. Federica Pirrone et al., "Owner-reported aggressive behavior towards familiar people may be a more prominent occurrence in pet shop-traded dogs," *Journal of Veterinary Behavior* 11 (2016): 13-17.

2. Meghan E. Herron, Frances S. Shofer, and Ilana R. Reisner, "Survey of the use and outcome of confrontational and non-confrontational training methods in client-owned dogs showing undesired behaviors," *Applied Animal Behaviour Science* 117, no. 1-2 (2009): 47-54.

3. Clare M. Browne et al., "Examination of the accuracy and applicability of information in popular books on dog training," *Society and Animals* 25, no. 5 (2017): 411-435.

4. Erica N. Feuerbacher and Clive D.L. Wynne, "Relative efficacy of human social interaction and food as reinforcers for domestic dogs and hand-reared wolves," *Journal of the Experimental Analysis of Behavior* 98, no. 1 (2012): 105-129.

5. Erica N. Feuerbacher and Clive D.L. Wynne, "Shut up and pet me! Domestic dogs (*Canis lupus familiaris*) prefer petting to vocal praise in concurrent and single-alternative choice procedures," *Behavioural Processes* 110 (2015): 47-59.

6. Erica N. Feuerbacher and Clive D.L. Wynne, "Most domestic dogs (*Canis lupus familiaris*) prefer food to petting: Population, context, and schedule effects in concurrent choice," *Journal of the Experimental Analysis of Behavior* 101, no. 3 (2014): 385-405.

7. Yuta Okamoto et al., "The feeding behavior of dogs correlates with their responses to commands," *Journal of Veterinary Medical Science* 71, no. 12 (2009): 1617-1621.

8. Megumi Fukuzawa and Naomi Hayashi, "Comparison of 3 different reinforcements of learning in dogs (*Canis familiaris*)," *Journal of Veterinary Behavior* 8, no. 4 (2013): 221–224.

9. Stefanie Riemer et al., "Reinforcer effectiveness in dogs—the influence of quantity and quality," *Applied Animal Behaviour Science* 206 (2018): 87–93.

10. Annika Bremhorst et al., "Incentive motivation in pet dogs—preference for constant vs varied food rewards," *Scientific Reports* 8, no. 1 (2018): 9756.

11. Cinzia Chiandetti et al., "Can clicker training facilitate conditioning in dogs?" *Applied Animal Behaviour Science* 184 (2016): 109–116.

12. Lynna C. Feng et al., "Is clicker training (clicker+food) better than food-only training for novice companion dogs and their owners?" *Applied Animal Behaviour Science* 204 (2018): 81–93.

13. Lynna C. Feng, Tiffani J. Howell, and Pauleen C. Bennett, "Practices and perceptions of clicker use in dog training: A survey-based investigation of dog owners and industry professionals," *Journal of Veterinary Behavior* 23 (2018): 1–9.

14. Clare M. Browne et al., "Delayed reinforcement—does it affect learning?" *Journal of Veterinary Behavior* 8, no. 4 (2013): e37–e38; Clare M. Browne et al., "Timing of reinforcement during dog training," *Journal of Veterinary Behavior* 6, no. 1 (2011): 58–59.

15. Nadja Affenzeller, Rupert Palme, and Helen Zulch, "Playful activity post-learning improves training performance in Labrador Retriever dogs (*Canis lupus familiaris*)," *Physiology & Behavior* 168 (2017): 62–73.

Chapter 5 The Vet and Grooming

1. John O. Volk et al., "Executive summary of the Bayer veterinary care usage study," *Journal of the American Veterinary Medical Association* 238, no. 10 (2011): 1275–1282.

2. C. Mariti et al., "The assessment of dog welfare in the waiting room of a veterinary clinic," *Animal Welfare* 24, no. 3 (2015): 299–305.

3. Chiara Mariti et al., "Guardians' perceptions of dogs' welfare and behaviors related to visiting the veterinary clinic," *Journal of Applied Animal Welfare Science* 20, no. 1 (2017): 24–33.

4. Marcy Hammerle et al., "2015 AAHA canine and feline behavior management guidelines," *Journal of the American Animal Hospital Association* 51, no. 4 (2015): 205–221.

5. Fear Free, "Fear Free veterinarians aim to reduce stress for pets," 2016, fearfreepets.com/fear-free-veterinarians-aim-to-reduce-stress-for-pets/.

6. Bruno Scalia, Daniela Alberghina, and Michele Panzera, "Influence of low stress handling during clinical visit on physiological and behavioural indicators in adult dogs: A preliminary study," *Pet Behaviour Science* 4 (2017): 20–22.

7. Karolina Westlund, "To feed or not to feed: Counterconditioning in the veterinary clinic," *Journal of Veterinary Behavior* 10, no. 5 (2015): 433–437.

8. Janice K.F. Lloyd, "Minimising stress for patients in the veterinary hospital: Why it is important and what can be done about it," *Veterinary Sciences* 4, no. 2 (2017): 22.

9. Erika Csoltova et al., "Behavioral and physiological reactions in dogs to a veterinary examination: Owner-dog interactions improve canine well-being," *Physiology & Behavior* 177 (2017): 270–281.

10. Rosalie Trevejo, Mingyin Yang, and Elizabeth M. Lund, "Epidemiology of surgical castration of dogs and cats in the United States," *Journal of the American Veterinary Medical Association* 238, no. 7 (2011): 898–904.

11. Margaret V. Root Kustritz et al., "Determining optimal age for gonad-ectomy in the dog: A critical review of the literature to guide decision making," *Journal of the American Veterinary Medical Association* 231, no. 11 (2007): 1665–1675.

12. Jessica M. Hoffman et al., "Do female dogs age differently than male dogs?" *The Journals of Gerontology: Series A* 73, no. 2 (2017): 150–156.

13. James A. Serpell and Yuying A. Hsu, "Effects of breed, sex, and neuter status on trainability in dogs," *Anthrozoös* 18, no. 3 (2005): 196–207.

14. Paul D. McGreevy et al., "Behavioural risks in male dogs with minimal lifetime exposure to gonadal hormones may complicate population-control benefits of desexing," *PLoS ONE* 13, no. 5 (2018): e0196284.

15. Paul D. McGreevy, Joanne Righetti, and Peter C. Thomson, "The reinforcing value of physical contact and the effect on canine heart rate of grooming in different anatomical areas," *Anthrozoös* 18, no. 3 (2005): 236–244.

16. Franziska Kuhne, Johanna C. Hößler, and Rainer Struwe, "Effects of human–dog familiarity on dogs' behavioural responses to petting," *Applied Animal Behaviour Science* 142, no. 3–4 (2012): 176–181.

17. Helen Vaterlaws-Whiteside and Amandine Hartmann, "Improv-ing puppy behavior using a new standardized socialization program," *Applied Animal Behaviour Science* 197 (2017): 55–61.

18. Franklin D. McMillan et al., "Differences in behavioral characteristics between dogs obtained as puppies from pet stores and those obtained from noncommercial breeders," *Journal of the American Veterinary Medical Association* 242, no. 10 (2013): 1359–1363.

19. Paul D. McGreevy et al., "Dog behavior co-varies with height, body-weight and skull shape," *PLoS ONE* 8, no. 12 (2013): e80529.

20. Todd W. Lue, Debbie P. Pantenburg, and Phillip M. Crawford, "Impact of the owner-pet and client-veterinarian bond on the care that pets receive," *Journal of the American Veterinary Medical Association* 232, no. 4 (2008): 531–540.

21. American Animal Hospital Association, "Frequency of veterinary visits," 2014, aaha.org/professional/resources/frequency_of_veterinary_visits. aspx.

22. Zoe Belshaw et al., "Owners and veterinary surgeons in the United Kingdom disagree about what should happen during a small animal vaccination consultation," *Veterinary Sciences* 5, no. 1 (2018): 7; Zoe Belshaw et al., "'I always feel like I have to rush . . .' Pet owner and small animal veterinary surgeons' reflections on time during preventative healthcare consultations in the United Kingdom," *Veterinary Sciences* 5, no. 1 (2018): 20.

23. Lawrence T. Glickman et al., "Evaluation of the risk of endocarditis and other cardiovascular events on the basis of the severity of periodontal disease in dogs," *Journal of the American Veterinary Medical Association* 234, no. 4 (2009): 486–494; Lawrence T. Glickman et al., "Association between chronic azotemic kidney disease and the severity of periodontal disease in dogs," *Preventive Veterinary Medicine* 99, no. 2–4 (2011): 193–200.

24. Steven E. Holmstrom et al., "2013 AAHA dental care guidelines for dogs and cats," *Journal of the American Animal Hospital Association* 49, no. 2 (2013): 75–82.

25. Judith L. Stella, Amy E. Bauer, and Candace C. Croney, "A cross-sectional study to estimate prevalence of periodontal disease in a population of dogs (*Canis familiaris*) in commercial breeding facilities in Indiana and Illinois," *PLoS ONE* 13, no. 1 (2018): e0191395.

Chapter 6 The Social Dog

1. Marc Bekoff, "Social play behavior. Cooperation, fairness, trust, and the evolution of morality," *Journal of Consciousness Studies* 8, no. 2 (2001): 81–90.

2. S.E. Byosiere, J. Espinosa, and B. Smuts, "Investigating the function of play bows in adult pet dogs (*Canis lupus familiaris*)," *Behavioural Processes* 125 (2016):106–113.

3. Sarah-Elizabeth Byosiere et al., "Investigating the function of play bows in dog and wolf puppies (*Canis lupus familiaris, Canis lupus occidentalis*)," *PLoS ONE* 11, no. 12 (2016): e0168570.

4. Alexandra Horowitz, "Attention to attention in domestic dog (*Canis familiaris*) dyadic play," *Animal Cognition* 12, no. 1 (2009): 107–118.

5. Marc Bekoff, "Play signals as punctuation: The structure of social play in canids," *Behaviour* (1995): 419–429.

6. Rebecca Sommerville, Emily A. O'Connor, and Lucy Asher, "Why do dogs play? Function and welfare implications of play in the domestic dog," *Applied Animal Behaviour Science* 197 (2017): 1–8.

7. Marek Spinka, Ruth C. Newberry, and Marc Bekoff, "Mammalian play: Training for the unexpected," *The Quarterly Review of Biology* 76, no. 2 (2001): 141–168.

8. Zsuzsánna Horváth, Antal Dóka, and Ádám Miklósi, "Affiliative and disciplinary behavior of human handlers during play with their dog affects cortisol concentrations in opposite directions," *Hormones and Behavior* 54, no. 1 (2008): 107–114.

9. Lydia Ottenheimer Carrier et al., "Exploring the dog park: Relationships between social behaviours, personality and cortisol in companion dogs," *Applied Animal Behaviour Science* 146, no. 1–4 (2013): 96–106.

10. Melissa S. Howse, Rita E. Anderson, and Carolyn J. Walsh, "Social behaviour of domestic dogs (*Canis familiaris*) in a public off-leash dog park," *Behavioural Processes* 157 (2018): 691–701.

11. John Bradshaw and Nicola Rooney, "Dog social behavior and communication," in *The Domestic Dog: Its Evolution, Behavior and Interactions with People*, ed. J. Serpell (Cambridge: Cambridge University Press, 2017), 133–159.

12. Neta-li Feuerstein and Joseph Terkel, "Interrelationships of dogs (*Canis familiaris*) and cats (*Felis catus L.*) living under the same roof," *Applied Animal Behaviour Science* 113, no. 1–3 (2008): 150–165.

13. Jessica E. Thomson, Sophie S. Hall, and Daniel S. Mills, "Evaluation of the relationship between cats and dogs living in the same home," *Journal of Veterinary Behavior* 27 (2018): 35–40.

14. Michael W. Fox, "Behavioral effects of rearing dogs with cats during the 'critical period of socialization,'" *Behaviour* 35, no. 3–4 (1969): 273–280.

Chapter 7 Dogs and Their People

1. Brian Hare and Michael Tomasello, "Human-like social skills in dogs?" *Trends in Cognitive Sciences* 9, no. 9 (2005): 439–444.

2. Juliane Kaminski, Andrea Pitsch, and Michael Tomasello, "Dogs steal in the dark," *Animal Cognition* 16, no. 3 (2013): 385–394; Juliane Bräuer

et al., "Domestic dogs conceal auditory but not visual information from others," *Animal Cognition* 16, no. 3 (2013): 351–359.

3. Charles H. Zeanah, Lisa J. Berlin, and Neil W. Boris, "Practitioner review: Clinical applications of attachment theory and research for infants and young children," *Journal of Child Psychology and Psychiatry* 52, no. 8 (2011): 819–833.

4. Elyssa Payne, Pauleen C. Bennett, and Paul D. McGreevy, "Current perspectives on attachment and bonding in the dog–human dyad," *Psychology Research and Behavior Management* 8 (2015): 71.

5. Márta Gácsi et al., "Human analogue safe haven effect of the owner: Behavioural and heart rate response to stressful social stimuli in dogs," *PLoS ONE* 8, no. 3 (2013): e58475.

6. Isabella Merola, Emanuela Prato-Previde, and Sarah Marshall-Pescini, "Social referencing in dog-owner dyads?" *Animal Cognition* 15, no. 2 (2012): 175–185.

7. Isabella Merola, Emanuela Prato-Previde, and Sarah Marshall-Pescini, "Dogs' social referencing towards owners and strangers," *PLoS ONE* 7, no. 10 (2012): e47653.

8. Erica N. Feuerbacher and Clive D.L. Wynne, "Dogs don't always prefer their owners and can quickly form strong preferences for certain strangers over others," *Journal of the Experimental Analysis of Behavior* 108, no. 3 (2017): 305–317.

9. Gregory S. Berns, Andrew M. Brooks, and Mark Spivak, "Scent of the familiar: An fMRI study of canine brain responses to familiar and unfamiliar human and dog odors," *Behavioural Processes* 110 (2015): 37–46.

10. Peter F. Cook et al., "Awake canine fMRI predicts dogs' preference for praise vs food," *Social Cognitive and Affective Neuroscience* 11, no. 12 (2016): 1853–1862.

11. Deborah Custance and Jennifer Mayer, "Empathic-like responding by domestic dogs (*Canis familiaris*) to distress in humans: An exploratory study," *Animal Cognition* 15, no. 5 (2012): 851–859.

12. Emily M. Sanford, Emma R. Burt, and Julia E. Meyers-Manor, "Timmy's in the well: Empathy and prosocial helping in dogs," *Learning & Behavior* 46, no. 4 (2018): 374–386.

13. Natalia Albuquerque et al., "Dogs recognize dog and human emotions," *Biology Letters* 12, no. 1 (2016): 20150883.

14. Hannah K. Worsley and Sean J. O'Hara, "Cross-species referential signalling events in domestic dogs (*Canis familiaris*)," *Animal Cognition* 21, no. 4 (2018): 457–465.

15. Nicola J. Rooney, John W.S. Bradshaw, and Ian H. Robinson, "A comparison of dog–dog and dog–human play behaviour," *Applied Animal Behaviour Science* 66, no. 3 (2000): 235–248.

16. Nicola J. Rooney, John W.S. Bradshaw, and Ian H. Robinson, "Do dogs respond to play signals given by humans?" *Animal Behaviour* 61, no. 4 (2001): 715–722.

17. Alexandra Horowitz and Julie Hecht, "Examining dog–human play: The characteristics, affect, and vocalizations of a unique interspecific interaction," *Animal Cognition* 19, no. 4 (2016): 779–788.

18. Tobey Ben-Aderet et al., "Dog-directed speech: Why do we use it and do dogs pay attention to it?" *Proceedings of the Royal Society B* 284, no. 1846 (2017): 20162429.

19. Sarah Jeannin et al., "Pet-directed speech draws adult dogs' attention more efficiently than adult-directed speech," *Scientific Reports* 7, no. 1 (2017): 4980.

20. Alex Benjamin and Katie Slocombe, "'Who's a good boy?!' Dogs prefer naturalistic dog-directed speech," *Animal Cognition* 21, no. 3 (2018): 353–364.

Chapter 8 Dogs and Children

1. Carri Westgarth et al., "Pet ownership, dog types and attachment to pets in 9–10 year old children in Liverpool, UK," *BMC Veterinary Research* 9, no. 1 (2013): 102.

2. Janine C. Muldoon, Joanne M. Williams, and Alistair Lawrence, "'Mum cleaned it and I just played with it': Children's perceptions of their roles and responsibilities in the care of family pets," *Childhood* 22, no. 2 (2015): 201–216.

3. Sophie S. Hall, Hannah F. Wright, and Daniel S. Mills, "Parent perceptions of the quality of life of pet dogs living with neuro-typically developing and neuro-atypically developing children: An exploratory study," *PLoS ONE* 12, no. 9 (2017): e0185300.

4. Nathaniel J. Hall et al., "Behavioral and self-report measures influencing children's reported attachment to their dog," *Anthrozoös* 29, no. 1 (2016): 137–150.

5. American Veterinary Medical Association (AVMA), "Dog bite prevention," Accessed March 31, 2018, avma.org/public/Pages/Dog-Bite-Prevention.aspx.

6. Ilana R. Reisner et al., "Behavioural characteristics associated with dog bites to children presenting to an urban trauma centre," *Injury Prevention* 17, no. 5 (2011): 348–353.

7. Yasemin Salgirli Demirbas et al., "Adults' ability to interpret canine body language during a dog–child interaction," *Anthrozoös* 29, no. 4 (2016): 581–596.

8. K. Meints, A. Racca, and N. Hickey, "How to prevent dog bite injuries? Children misinterpret dogs facial expressions," *Injury Prevention* 16, Suppl 1 (2010): A68.

9. Christine Arhant, Andrea Martina Beetz, and Josef Troxler, "Caregiver reports of interactions between children up to 6 years and their family dog—implications for dog bite prevention," *Frontiers in Veterinary Science* 4 (2017): 130.

10. Christine Arhant et al., "Attitudes of caregivers to supervision of child-family dog interactions in children up to 6 years—an exploratory study," *Journal of Veterinary Behavior* 14 (2016): 10–16.

11. Jiabin Shen et al., "Systematic review: Interventions to educate children about dog safety and prevent pediatric dog-bite injuries: A meta-analytic review," *Journal of Pediatric Psychology* 42, no. 7 (2016): 779–791.

12. Sato Arai, Nobuyo Ohtani, and Mitsuaki Ohta, "Importance of bringing dogs in contact with children during their socialization period for better behavior," *Journal of Veterinary Medical Science* 73, no. 6 (2011): 747–752.

13. Carlo Siracusa, Lena Provoost, and Ilana R. Reisner, "Dog- and owner-related risk factors for consideration of euthanasia or rehoming before a referral behavioral consultation and for euthanizing or rehoming the dog after the consultation," *Journal of Veterinary Behavior* 22 (2017): 46–56.

Chapter 9 Time for Walkies!

1. Dawn Brooks et al., "2014 AAHA weight management guidelines for dogs and cats," *Journal of the American Animal Hospital Association* 50, no. 1 (2014): 1–11.

2. C.A. Pugh et al., "Dogslife: A cohort study of Labrador Retrievers in the UK," *Preventive Veterinary Medicine* 122, no. 4 (2015): 426–435.

3. Sarah E. Lofgren et al., "Management and personality in Labrador Retriever dogs," *Applied Animal Behaviour Science* 156 (2014): 44–53.

4. Tiffani J. Howell, Kate Mornement, and Pauleen C. Bennett, "Pet dog management practices among a representative sample of owners in Victoria, Australia," *Journal of Veterinary Behavior* 12 (2016): 4–12.

5. Carri Westgarth et al., "Dog behavior on walks and the effect of use of the leash," *Applied Animal Behaviour Science* 125, no. 1–2 (2010): 38–46.

6. Carri Westgarth et al., "I walk my dog because it makes me happy: A qualitative study to understand why dogs motivate walking and improved health," *International Journal of Environmental Research and Public Health* 14, no. 8 (2017): 936.

7. Chris Degeling and Melanie Rock, "'It was not just a walking experience': Reflections on the role of care in dog-walking," *Health Promotion International* 28, no. 3 (2012): 397–406.

8. Christine Arhant et al., "Behaviour of smaller and larger dogs: Effects of training methods, inconsistency of owner behaviour and level of engagement in activities with the dog," *Applied Animal Behaviour Science* 123, no. 3–4 (2010): 131–142.

9. Chris Degeling, Lindsay Burton, and Gavin R. McCormack, "An investigation of the association between socio-demographic factors, dog-exercise requirements, and the amount of walking dogs receive," *Canadian Journal of Veterinary Research* 76, no. 3 (2012): 235–240.

10. Hayley Christian et al., "Encouraging dog walking for health promotion and disease prevention," *American Journal of Lifestyle Medicine* 12, no. 3 (2018): 233–243.

11. Amanda Jane Kobelt et al., "The behaviour of Labrador Retrievers in subur-
    ban backyards: The relationships between the backyard environment and
    dog behaviour," *Applied Animal Behaviour Science* 106, no. 1–3 (2007): 70–84.

12. Westgarth et al., "Dog behavior on walks."

13. Rachel Moxon, H. Whiteside, and Gary C.W. England, "Incidence and
    impact of dog attacks on guide dogs in the UK: An update," *Veterinary
    Record* 178, no. 15 (2016): 367.

14. San Francisco Society for the Prevention of Cruelty to Animals (SF SPCA),
    "Prong collars: Myths and facts," Accessed March 31, 2018, sfspca.org/
    prong/myths.

15. John Grainger, Alison P. Wills, and V. Tamara Montrose, "The behavioral
    effects of walking on a collar and harness in domestic dogs (*Canis famil-
    iaris*)," *Journal of Veterinary Behavior* 14 (2016): 60–64.

Chapter 10 Enrichment

1. Nicola J. Rooney and John W.S. Bradshaw, "An experimental study of
   the effects of play upon the dog–human relationship," *Applied Animal
   Behaviour Science* 75, no. 2 (2002): 161–176.

2. Ragen T.S. McGowan et al., "Positive affect and learning: Exploring the
   'Eureka Effect' in dogs," *Animal Cognition* 17, no. 3 (2014): 577–587.

3. Christine Arhant et al., "Behavior of smaller and larger dogs: Effects
   of training methods, inconsistency of owner behaviour and level of
   engagement in activities with the dog," *Applied Animal Behaviour
   Science* 123, no. 3–4 (2010): 131–142.

4. John Bradshaw and Nicola Rooney, "Dog social behavior and communication" in *The Domestic Dog: Its Evolution, Behavior and Interactions with*

People, ed. James Serpell (Cambridge: Cambridge University Press, 2017), 133–159.

5. George M. Strain, "How well do dogs and other animals hear?" Accessed March 31, 2018, lsu.edu/deafness/HearingRange.html.

6. Lori R. Kogan, Regina Schoenfeld-Tacher, and Allen A. Simon, "Behavioral effects of auditory stimulation on kenneled dogs," *Journal of Veterinary Behavior* 7, no. 5 (2012): 268–275.

7. A. Bowman et al., "'Four Seasons' in an animal rescue centre; classical music reduces environmental stress in kennelled dogs," *Physiology & Behavior* 143 (2015): 70–82.

8. Alexandra A. Horowitz, *Being a Dog: Following the Dog into a World of Smell* (New York: Scribner, 2016).

9. C. Duranton and A. Horowitz, "Let me sniff! Nosework induces positive judgment bias in pet dogs," *Applied Animal Behaviour Science* 211 (2019): 61–66.

10. Jocelyn (Joey) M. Farrell et al., "Dog-sport competitors: What motivates people to participate with their dogs in sporting events?" *Anthrozoös* 28, no. 1 (2015): 61–71.

11. Camilla Pastore et al., "Evaluation of physiological and behavioral stress-dependent parameters in agility dogs," *Journal of Veterinary Behavior* 6, no. 3 (2011): 188–194.

12. Anne J. Pullen, Ralph J.N. Merrill, and John W.S. Bradshaw, "Habituation and dishabituation during object play in kennel-housed dogs," *Animal Cognition* 15, no. 6 (2012): 1143–1150.

13. Lidewij L. Schipper et al., "The effect of feeding enrichment toys on the behavior of kennelled dogs (*Canis familiaris*)," *Applied Animal Behaviour Science* 114, no. 1–2 (2008): 182–195.

14. Jenna Kiddie and Lisa Collins, "Identifying environmental and management factors that may be associated with the quality of life of kennelled dogs (*Canis familiaris*)," *Applied Animal Behaviour Science* 167 (2015): 43–55.

Chapter 11 Food and Treats

1. Erik Axelsson et al., "The genomic signature of dog domestication reveals adaptation to a starch-rich diet," *Nature* 495, no. 7441 (2013): 360.

2. Maja Arendt et al., "Diet adaptation in dog reflects spread of prehistoric agriculture," *Heredity* 117, no. 5 (2016): 301.

3. Maja Arendt et al., "Amylase activity is associated with AMY 2B copy numbers in dog: Implications for dog domestication, diet and diabetes," *Animal Genetics* 45, no. 5 (2014): 716–722.

4. Morgane Ollivier et al., "Amy2B copy number variation reveals starch diet adaptations in ancient European dogs," *Royal Society Open Science* 3, no. 11 (2016): 160449.

5. Tiffani J. Howell, Kate Mornement, and Pauleen C. Bennett, "Pet dog management practices among a representative sample of owners in Victoria, Australia," *Journal of Veterinary Behavior* 12 (2016): 4–12.

6. C.A. Pugh et al., "Dogslife: A cohort study of Labrador Retrievers in the UK," *Preventive Veterinary Medicine* 122, no. 4 (2015): 426–435.

7. Kathryn E. Michel, "Unconventional diets for dogs and cats," *Veterinary Clinics: Small Animal Practice* 36, no. 6 (2006): 1269–1281.

8. Vivian Pedrinelli, Márcia de O.S. Gomes, and Aulus C. Carciofi, "Analysis of recipes of home-prepared diets for dogs and cats published in Portuguese," *Journal of Nutritional Science* 6 (2017): e33.

9. Andrew Knight and Madelaine Leitsberger, "Vegetarian versus meat-based diets for companion animals," *Animals* 6, no. 9 (2016): 57.

10. Daniel P. Schlesinger and Daniel J. Joffe, "Raw food diets in companion animals: A critical review," *The Canadian Veterinary Journal* 52, no. 1 (2011): 50.

11. Freek P.J. van Bree et al., "Zoonotic bacteria and parasites found in raw meat-based diets for cats and dogs," *Veterinary Record* 182, no. 2 (2018): 50.

12. J. Boyd, "Should you feed your pet raw meat? The risks of a 'traditional' diet," 2018, phys.org/news/2018-01-pet-raw-meat-real-traditional.html.

13. ASPCA Poison Control, "People foods to avoid feeding your pet," Accessed September 30, 2018, aspca.org/pet-care/animal-poison-control/people-foods-avoid-feeding-your-pets.

14. Giada Morelli et al., "Study of ingredients and nutrient composition of commercially available treats for dogs," *Veterinary Record* 182, no. 12 (2018): 351.

15. Ernie Ward, Alexander J. German, and Julie A. Churchill, "The Global Pet Obesity Initiative position statement," Accessed December 29, 2018, petobesityprevention.org/about.

16. Elizabeth M. Lund et al., "Prevalence and risk factors for obesity in adult dogs from private US veterinary practices," *International Journal of Applied Research in Veterinary Medicine* 4, no. 2 (2006): 177; P.D. McGreevy et al., "Prevalence of obesity in dogs examined by Australian veterinary practices and the risk factors involved," *Veterinary Record—English Edition* 156, no. 22 (2005): 695–701.

17. Alexander J. German et al., "Small animal health: Dangerous trends in pet obesity," *Veterinary Record* 182, no. 1 (2018): 25.

18. Ellen Kienzle, Reinhold Bergler, and Anja Mandernach, "A comparison of the feeding behavior and the human–animal relationship in owners of normal and obese dogs," *The Journal of Nutrition* 128, no. 12 (1998): 2779S–2782S.

19. Vanessa I. Rohlf et al., "Dog obesity: Can dog caregivers' (owners') feeding and exercise intentions and behaviors be predicted from attitudes?" *Journal of Applied Animal Welfare Science* 13, no. 3 (2010): 213–236.

20. Marta Krasuska and Thomas L. Webb, "How effective are interventions designed to help owners to change their behaviour so as to manage the weight of their companion dogs? A systematic review and meta-analysis," *Preventive Veterinary Medicine* 159, no. 1 (2018): 40–50.

21. Carina Salt et al., "Association between life span and body condition in neutered client-owned dogs," *Journal of Veterinary Internal Medicine* 33, no. 1 (2018): 89–99.

22. Banfield Veterinary Hospital, "Obesity is an epidemic," Accessed September 29, 2018, banfield.com/state-of-pet-health/obesity.

23. Jaak Panksepp and Margaret R. Zellner, "Towards a neurobiologically based unified theory of aggression," *Revue internationale de psychologie sociale* 17 (2004): 37–62.

24. Ray Coppinger and L. Coppinger, *Dogs: A Startling New Understanding of Canine Origin, Behavior and Evolution* (New York: Scribner, 2001).

25. Monique A.R. Udell et al., "Exploring breed differences in dogs (*Canis familiaris*): Does exaggeration or inhibition of predatory response predict performance on human-guided tasks?," *Animal Behaviour* 89 (2014): 99–105; D. Horwitz, *Blackwell's Five-Minute Veterinary Consult Clinical*

Companion: Canine and Feline Behavior, 2nd edition (Oxford, UK: Wiley Blackwell, 2017).

Chapter 12 Sleeping Dogs

1. C.A. Pugh et al., "Dogslife: A cohort study of Labrador Retrievers in the UK," *Preventive Veterinary Medicine* 122, no. 4 (2015): 426–435.

2. Tiffani J. Howell, Kate Mornement, and Pauleen C. Bennett, "Pet dog management practices among a representative sample of owners in Victoria, Australia," *Journal of Veterinary Behavior* 12 (2016): 4–12.

3. Victoria L. Voith, John C. Wright, and Peggy J. Danneman, "Is there a relationship between canine behavior problems and spoiling activities, anthropomorphism and obedience training?" *Applied Animal Behaviour Science* 34, no. 3 (1992): 263–272.

4. Bradley Smith et al., "The prevalence and implications of human–animal co-sleeping in an Australian sample," *Anthrozoös* 27, no. 4 (2014): 543–551.

5. Christy L. Hoffman, Kaylee Stutz, and Terrie Vasilopoulos, "An examination of adult women's sleep quality and sleep routines in relation to pet ownership and bedsharing," *Anthrozoös* 31, no. 6 (2018): 711–725.

6. Simona Cannas et al., "Factors associated with dog behavioral problems referred to a behavior clinic," *Journal of Veterinary Behavior* 24 (2018): 42–47.

7. Dorothea Döring et al., "Use of beds by laboratory beagles," *Journal of Veterinary Behavior* 28 (2018): 6–10.

8. Dorothea Döring et al., "Behavioral observations in dogs in four research facilities: Do they use their enrichment?" *Journal of Veterinary Behavior* 13 (2016): 55–62.

9. Brian M. Zanghi et al., "Effect of age and feeding schedule on diurnal rest/activity rhythms in dogs," *Journal of Veterinary Behavior* 7, no. 6 (2012): 339–347.

10. Scott S. Campbell and Irene Tobler, "Animal sleep: A review of sleep duration across phylogeny," *Neuroscience & Biobehavioral Reviews* 8, no. 3 (1984): 269–300.

11. Sara C. Owczarczak-Garstecka and Oliver H.P. Burman, "Can sleep and resting behaviours be used as indicators of welfare in shelter dogs (*Canis lupus familiaris*)?" *PLoS ONE* 11, no. 10 (2016): e0163620.

12. Zanghi, "Effect of age and feeding schedule."

13. Brian M. Zanghi et al., "Characterizing behavioral sleep using actigraphy in adult dogs of various ages fed once or twice daily," *Journal of Veterinary Behavior* 8, no. 4 (2013): 195–203.

14. R. Fast et al., "An observational study with long-term follow-up of canine cognitive dysfunction: Clinical characteristics, survival, and risk factors," *Journal of Veterinary Internal Medicine* 27, no. 4 (2013): 822–829.

15. G.J. Adams and K.G. Johnson, "Behavioral responses to barking and other auditory stimuli during night-time sleeping and waking in the domestic dog (*Canis familiaris*)," *Applied Animal Behaviour Science* 39, no. 2 (1994): 151–162.

16. G.J. Adams and K.G. Johnson, "Sleep-wake cycles and other night-time behaviors of the domestic dog *Canis familiaris*," *Applied Animal Behaviour Science* 36, no. 2 (1993): 233–248.

17. Anna Kis et al., "Sleep macrostructure is modulated by positive and negative social experience in adult pet dogs," *Proceedings of the Royal Society B* 284, no. 1865 (2017): 20171883.

18. Anna Kis et al., "The interrelated effect of sleep and learning in dogs (*Canis familiaris*); an EEG and behavioural study," *Scientific Reports* 7 (2017): 41873.

19. Helle Demant et al., "The effect of frequency and duration of training sessions on acquisition and long-term memory in dogs," *Applied Animal Behaviour Science* 133, no. 3–4 (2011): 228–234.

Chapter 13 Fear and Other Problems

1. Niwako Ogata, "Separation anxiety in dogs: What progress has been made in our understanding of the most common behavioral problems in dogs?" *Journal of Veterinary Behavior* 16 (2016): 28–35.

2. E.L. Buckland et al., "Prioritisation of companion dog welfare issues using expert consensus," *Animal Welfare* 23, no. 1 (2014): 39–46.

3. Alexandra Horowitz, "Disambiguating the 'guilty look': Salient prompts to a familiar dog behaviour," *Behavioural Processes* 81, no. 3 (2009): 447–452.

4. Julie Hecht, Ádám Miklósi, and Márta Gácsi, "Behavioral assessment and owner perceptions of behaviors associated with guilt in dogs," *Applied Animal Behaviour Science* 139, no. 1–2 (2012): 134–142.

5. James A. Serpell and Deborah L. Duffy, "Aspects of juvenile and adolescent environment predict aggression and fear in 12-month-old guide dogs," *Frontiers in Veterinary Science* 3 (2016): 49.

6. Isain Zapata, James A. Serpell, and Carlos E. Alvarez, "Genetic mapping of canine fear and aggression," *BMC Genomics* 17, no. 1 (2016): 572.

7. Moshe Szyf, "DNA methylation, behavior and early life adversity," *Journal of Genetics and Genomics* 40, no. 7 (2013): 331–338; Jana P. Lim and Anne Brunet, "Bridging the transgenerational gap with epigenetic memory," *Trends in Genetics* 29, no. 3 (2013): 176–186.

8. Patricia Vetula Gallo, Jack Werboff, and Kirvin Knox, "Development of home orientation in offspring of protein-restricted cats," *Developmental Psychobiology: The Journal of the International Society for Developmental Psychobiology* 17, no. 5 (1984): 437–449.

9. Pernilla Foyer, Erik Wilsson, and Per Jensen, "Levels of maternal care in dogs affect adult offspring temperament," *Scientific Reports* 6 (2016): 19253.

10. Giovanna Guardini et al., "Influence of morning maternal care on the behavioural responses of 8-week-old Beagle puppies to new environmental and social stimuli," *Applied Animal Behaviour Science* 181 (2016): 137–144.

11. Lisa Jessica Wallis et al., "Demographic change across the lifespan of pet dogs and their impact on health status," *Frontiers in Veterinary Science* 5 (2018): 200.

12. Michele Wan, Niall Bolger, and Frances A. Champagne, "Human perception of fear in dogs varies according to experience with dogs," *PLoS ONE* 7, no. 12 (2012): e51775.

13. Emily J. Blackwell, John W.S. Bradshaw, and Rachel A. Casey, "Fear responses to noises in domestic dogs: Prevalence, risk factors and co-occurrence with other fear related behaviour," *Applied Animal Behaviour Science* 145, no. 1–2 (2013): 15–25.

14. Rachel A. Casey et al., "Human directed aggression in domestic dogs (*Canis familiaris*): Occurrence in different contexts and risk factors," *Applied Animal Behaviour Science* 152 (2014): 52–63.

15. Animal Legal and Historical Center, "Ontario Statutes—Dog Owners' Liability Act," Accessed March 16, 2019, animallaw.info/statute/canada-ontario-dog-owners-liability-act.

16. L.S. Weiss, "Breed-specific legislation in the United States," Animal Legal and Historical Center, 2001, animallaw.info/article/breed-specific-legislation-united-states.

17. Royal Society for the Prevention of Cruelty to Animals (RSPCA), "Breed specific legislation, a dog's dinner," 2016, rspca.org.uk/webContent/staticImages/Downloads/BSL_Report.pdf.

18. Páraic Ó Súilleabháin, "Human hospitalisations due to dog bites in Ireland (1998–2013): Implications for current breed specific legislation," *The Veterinary Journal* 204, no. 3 (2015): 357–359.

19. Finn Nilson et al., "The effect of breed-specific dog legislation on hospital treated dog bites in Odense, Denmark—a time series intervention study," *PLoS ONE* 13, no. 12 (2018): e0208393.

20. Christy L. Hoffman et al., "Is that dog a pit bull? A cross-country comparison of perceptions of shelter workers regarding breed identification," *Journal of Applied Animal Welfare Science* 17, no. 4 (2014): 322–339.

21. City of Calgary, "Bylaws related to dogs," Accessed March 31, 2018, calgary.ca/CSPS/ABS/Pages/Bylaws-by-topic/Dogs.aspx.

22. René Bruemmer, "How Calgary reduced dog attacks without banning pit bulls," *Montreal Gazette,* September 1, 2016.

23. Carri Westgarth and Francine Watkins, "A qualitative investigation of the perceptions of female dog-bite victims and implications for the prevention of dog bites," *Journal of Veterinary Behavior* 10, no. 6 (2015): 479–488.

24. Nicole S. Starinsky, Linda K. Lord, and Meghan E. Herron, "Escape rates and biting histories of dogs confined to their owner's property through the use of various containment methods," *Journal of the American Veterinary Medical Association* 250, no. 3 (2017): 297–302.

25. Meghan E. Herron, Frances S. Shofer, and Ilana R. Reisner, "Survey of the use and outcome of confrontational and non-confrontational training methods in client-owned dogs showing undesired behaviors," *Applied Animal Behaviour Science* 117, no. 1–2 (2009): 47–54.

26. Karen Overall, *Manual of Clinical Behavioral Medicine for Dogs and Cats* (St. Louis, MO: Elsevier Health Sciences, 2013).

27. E. Blackwell, R.A. Casey, and J.W.S. Bradshaw, "Controlled trial of behavioural therapy for separation-related disorders in dogs," *Veterinary Record* 158, no. 16 (2006): 551–554.

28. Malena DeMartini-Price, *Treating Separation Anxiety in Dogs* (Wenatchee, WA: Dogwise Publishing, 2014).

29. Jacquelyn A. Jacobs et al., "Ability of owners to identify resource guarding behaviour in the domestic dog," *Applied Animal Behaviour Science* 188 (2017): 77–83.

30. Heather Mohan-Gibbons, Emily Weiss, and Margaret Slater, "Preliminary investigation of food guarding behavior in shelter dogs in the United States," *Animals* 2, no. 3 (2012): 331–346; Amy R. Marder et al., "Food-related aggression in shelter dogs: A comparison of behavior identified by a behavior evaluation in the shelter and owner reports after adoption," *Applied Animal Behaviour Science* 148, no. 1–2 (2013): 150–156.

31. Jacquelyn A. Jacobs et al., "Factors associated with canine resource guarding behaviour in the presence of people: A cross-sectional survey of dog owners," *Preventive Veterinary Medicine* 161 (2018): 143–153.

32. Overall, *Manual of Clinical Behavioral Medicine for Dogs and Cats.*

33. Ana Luisa Lopes Fagundes et al., "Noise sensitivities in dogs: An exploration of signs in dogs with and without musculoskeletal pain using qualitative content analysis," *Frontiers in Veterinary Science* 5 (2018): 17.

34. Carlo Siracusa, Lena Provoost, and Ilana R. Reisner, "Dog- and owner-related risk factors for consideration of euthanasia or rehoming before a referral behavioral consultation and for euthanizing or rehoming the dog after the consultation," *Journal of Veterinary Behavior* 22 (2017): 46–56.

Chapter 14 Seniors and Dogs with Special Needs

1. Lisa Jessica Wallis et al., "Lifespan development of attentiveness in domestic dogs: Drawing parallels with humans," *Frontiers in Psychology* 5 (2014): 71.

2. Jan Bellows et al., "Common physical and functional changes associated with aging in dogs," *Journal of the American Veterinary Medical Association* 246, no. 1 (2015): 67–75; Hannah E. Salvin et al., "The effect of breed on age-related changes in behavior and disease prevalence in cognitively normal older community dogs, *Canis lupus familiaris*," *Journal of Veterinary Behavior* 7, no. 2 (2012): 61–69.

3. Naomi Harvey, "Imagining life without Dreamer," *Veterinary Record* 182 (2018): 299.

4. Durga Chapagain et al., "Aging of attentiveness in Border Collies and other pet dog breeds: The protective benefits of lifelong training," *Frontiers in Aging Neuroscience* 9 (2017): 100.

5. Elizabeth Head, "Combining an antioxidant-fortified diet with behavioral enrichment leads to cognitive improvement and reduced brain pathology in aging canines: Strategies for healthy aging," *Annals of the New York Academy of Sciences* 1114, no. 1 (2007): 398–406.

6. Yuanlong Pan et al., "Cognitive enhancement in old dogs from dietary supplementation with a nutrient blend containing arginine, antioxidants, B vitamins and fish oil," *British Journal of Nutrition* 119, no. 3 (2018): 349–358.

7. Valeri Farmer-Dougan et al., "Behavior of hearing or vision impaired and normal hearing and vision dogs (*Canis lupis familiaris*): Not the same, but not that different," *Journal of Veterinary Behavior* 9, no. 6 (2014): 316–323.

8. J. Kirpensteijn, R. Van, and N. Endenburg Bos, "Adaptation of dogs to the amputation of a limb and their owners' satisfaction with the procedure," *Veterinary Record* 144, no. 5 (1999): 115–118.

Chapter 15 The End of Life

1. Mai Inoue, "A current life table and causes of death for insured dogs in Japan," *Preventive Veterinary Medicine* 120, no. 2 (2015): 210–218.

2. Inoue, "Current life table."

3. V.J. Adams et al., "Methods and mortality results of a health survey of purebred dogs in the UK," *Journal of Small Animal Practice* 51, no. 10 (2010): 512–524.

4. J.M. Fleming, K.E. Creevy, and D.E.L. Promislow, "Mortality in North American dogs from 1984 to 2004: An investigation into age-, size-, and breed-related causes of death," *Journal of Veterinary Internal Medicine* 25, no. 2 (2011): 187–198.

5. D.G. O'Neill et al., "Longevity and mortality of owned dogs in England," *The Veterinary Journal* 198, no. 3 (2013): 638–643.

6. Peter Sandøe, Clare Palmer, and Sandra Corr, "Human attachment to dogs and cats and its ethical implications," in *22nd FECAVA Eurocongress* 31 (2016): 11–14.

7. Belshaw et al., "Quality of life assessment in domestic dogs: An evidence-based rapid review," *The Veterinary Journal* 206, no. 2 (2015): 203–212.

8. Alice Villalobos, "Cancers in dogs and cats," in *Hospice and Palliative Care for Companion Animals: Principles and Practice*, eds. A. Shanan, T. Shearer, and J. Pierce (Hoboken: Wiley-Blackwell, 2017): 89-100.

9. Vivian C. Fan et al., "Retrospective survey of owners' experiences with palliative radiation therapy for pets," *Journal of the American Veterinary Medical Association* 253, no. 3 (2018): 307-314.

10. Stine Billeschou Christiansen et al., "Veterinarians' role in clients' decision-making regarding seriously ill companion animal patients," *Acta Veterinaria Scandinavica* 58, no. 1 (2015): 30.

11. Sandra Barnard-Nguyen et al., "Pet loss and grief: Identifying at-risk pet owners during the euthanasia process," *Anthrozoös* 29, no. 3 (2016): 421-430.

12. Lilian Tzivian, Michael Friger, and Talma Kushnir, "Associations between stress and quality of life: Differences between owners keeping a living dog or losing a dog by euthanasia," *PLoS ONE* 10, no. 3 (2015): e0121081.

13. Jessica K. Walker, Natalie K. Waran, and Clive J.C. Phillips, "Owners' perceptions of their animal's behavioral response to the loss of an animal companion," *Animals* 6, no. 11 (2016): 68.

14. Sakiko, Yamazaki, "A survey of companion-animal owners affected by the East Japan Great Earthquake in Iwate and Fukushima Prefectures, Japan," *Anthrozoös* 28, no. 2 (2015): 291-304.

ACKNOWLEDGMENTS

M ANY PEOPLE HAVE been supportive of this book and I am grateful to them all. Thank you to everyone who spoke to me or answered questions by email and took the time to think about their research from a dog's perspective.

Jean Donaldson is the best dog training mentor anyone could ask for and I draw on the Academy's thorough education in dog training all the time.

If I did not blog, I would not have thought to write this book, and everyone who has liked, shared, or left a nice comment on *Companion Animal Psychology* has encouraged me to keep going. Thank you in particular to Marc Bekoff, Mia Cobb, Mikel Delgado, Julie Hecht, Jessica Hekman, Hal Herzog, Kat Littlewood, and Kate Mornement for setting a high bar for science blogging, to Eileen Anderson for her dog training blog (anyone who does not follow their blogs should do!), and to Malcolm M. Campbell for all the puns and shares. I very much appreciate the team at Science Borealis for all they do to support Canadian science blogging.

Special thanks to Kristi Benson, Sylvie Martin, and Beth Sautins for their friendship and helpful comments on earlier versions of some chapters. I am grateful to Suzanne Bryner, Nickala Squire,

Cara Moynes, Nick Honor, Joan Hunter-Mayer, Tim Steele, Kathrine Mancuso Christ, Julie Parker, Jody Karow, Jenn Bauer, Joan Grassbaugh Forry, Kayla Block, Lori Nanan, Melanie Diantoniis Cerone, Rachel Szumel, Linda Green, Erica Beckwith, Claudine Prud'homme, and everyone in the Academy writing group for all the encouragement and understanding along the way. Kind words from Kathy Sdao have meant more than you can imagine. And thank you to Roy and Frankie Todd, Stef Harvey, Bonnie Hartney, Helen Verte, Tracy Krulik, Kim Monteith, Eva Kifri, Nathalie Mosbach Smith, Corey Van't Haaff, José Kahan Oblatt, and all of the walking group for their support.

Rummy Evans of Bad Monkey Photography, Jean Ballard, Kristy Francis, and Christine Michaud have kindly let me use their wonderful dog photos.

I have learned from all the dogs I've hung out with and/or trained over the years, but in particular Ghost, Bodger, Charlie, Rex, Burton, Marshall, Tess, Johnny Ombré, and Junior.

I will always be grateful to my first agent, Trena White, and my new agent, Fiona Kenshole, for their invaluable advice. Thank you to everyone at Greystone Books who has worked so hard to make this book a reality and helped ensure it is the best it can be. I'm especially grateful to my editor, Lucy Kenward, for her curiosity, patience, and guidance; the book is much better for her input. Thank you to my copy editor, Rowena Rae, for bringing consistency and clarity to the manuscript.

Special thanks to Al for his support and encouragement all the way through.

It goes without saying that all mistakes and omissions are my own.

INDEX

Illustrations and tables indicated by page numbers in italics